GW01158371

How to solve the world's Mathematical Olympiad problems

Volume I

Steve Dinh, a.k.a. Vo Duc Dien

AuthorHouse™
1663 Liberty Drive
Bloomington, IN 47403
www.authorhouse.com
Phone: 1-800-839-8640

First published by AuthorHouse 11/16/2010

ISBN: 978-1-4520-5177-2 (sc)
ISBN: 978-1-4520-5178-9 (e-b)

Printed in the United States of America

This book is printed on acid-free paper.

To my wife Quynh-Chau,
to my children Catherine Diem and Alan Huy,
to my mother Thuy,
and to my dad in heaven, De Van

Problem 1 of the International Mathematical Olympiad 2006

Let ABC be a triangle with incenter I. A point P in the interior of the triangle satisfies $\angle PBA + \angle PCA = \angle PBC + \angle PCB$. Show that $AP > AI$, and that equality holds if and only if $P = I$.

Solution

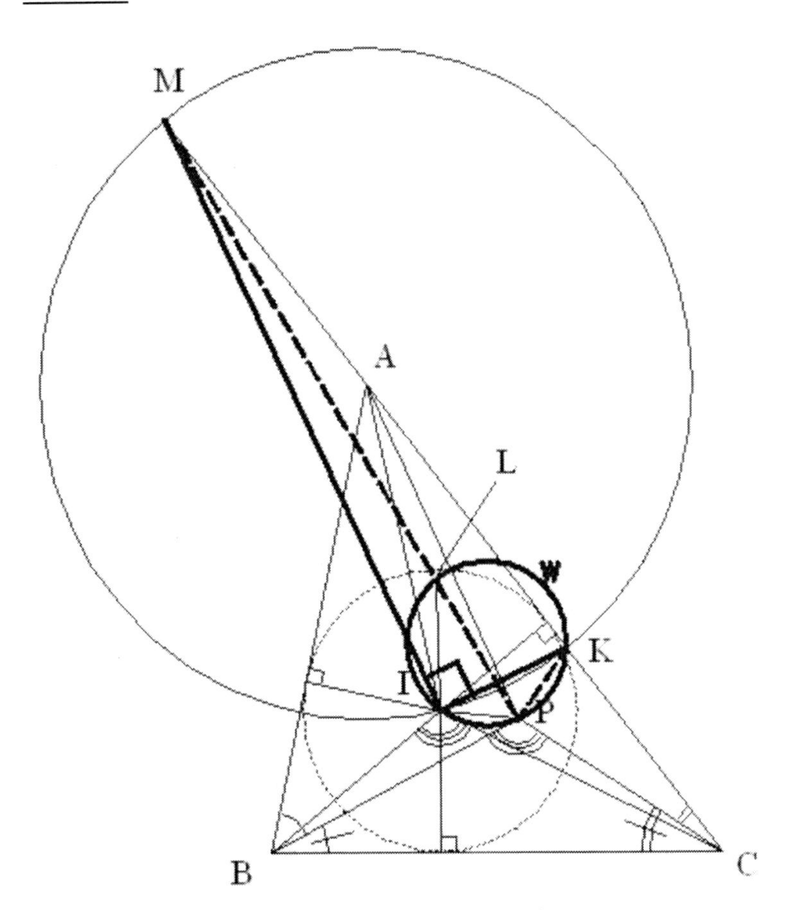

We have $\angle BIC = 180 - \frac{1}{2}(\angle ABC + \angle ACB)$

$\angle BPC = 180 - (\angle PBC + \angle PCB)$

The problem gives us
$\angle PBA + \angle PCA = \angle PBC + \angle PCB = \frac{1}{2}(\angle ABC + \angle ACB)$

Therefore, $\angle BPC = \angle BIC$.

Draw a circle with center A that passes through I and intersects AC at K.

We have

$$\angle MPK = 360° - \angle BPC - \angle MPB - \angle KPC \qquad \text{(i)}$$

But $\angle BPC = \angle BIC$

$\angle MPB = \angle MIB - \angle IMP - \angle IBP$

$\angle KPC = \angle KIC + \angle IKP + \angle ICP$

$\angle IBP = \frac{1}{2} \angle ABC - \angle PBC$

$\angle ICP = \angle PCB - \frac{1}{2} \angle ACB$

and (i) becomes $\angle MPK = 360° - \angle BIC - \angle MIB + \angle IMP + \angle IBP - \angle KIC - \angle IKP - \angle ICP = 360° - \angle BIC - \angle MIB + \angle IMP + \frac{1}{2}\angle ABC - \angle PBC - \angle KIC - \angle IKP + \frac{1}{2} \angle ACB - \angle PCB$

Since $\frac{1}{2} (\angle ABC + \angle ACB) = \angle PBC + \angle PCB$

$\angle MPK = (360° - \angle BIC - \angle MIB - \angle KIC) + \angle IMP - \angle IKP$

$\angle MIK = 360° - \angle BIC - \angle MIB - \angle KIC = 90°$

$\angle MPK = 90° + \angle IMP - \angle IKP$

$= 90° + \angle IMP - \angle ILP$

Also since $\angle ILP > \angle IMP$

$\angle IMP - \angle ILP < 0$

or $\angle MPK < 90°$.

Therefore, point P is outside the circle with center A and

AP > AI.

Problem 1 of the Asian Pacific Mathematical Olympiad 1991

Let G be the centroid of triangle ABC and M be the midpoint of BC. Let X be on AB and Y on AC such that the points X, Y, and G are collinear and XY and BC are parallel. Suppose that XC and GB intersect at Q and YB and GC intersect at P. Show that triangle MPQ is similar to triangle ABC.

Solution

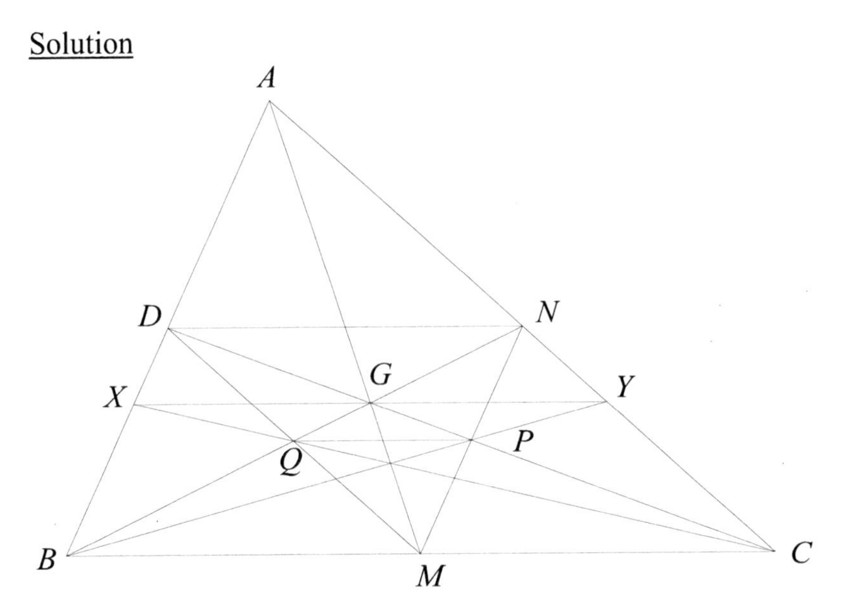

Let N and D be midpoints of AC and AB, respectively.
Since XY || BC, we have YC / YN = GB / GN, or
(YC/YN) × (GN/GB) = 1. Since M is midpoint of BC, we have
(MB/MC) × (YC/YN) × (GN/GB) = 1

Per Ceva's theorem, the three lines MN, GC and BY are concurrent and meet at Q. G is the centroid of triangle ABC, link CD. Since MN || AB and D is midpoint of AB, P is then midpoint of MN. We have MP || AB.
Using the same argument, Q is midpoint of MD and MQ || AC and PQ || DN.
In addition with DN || BC, we have PQ || BC. Triangle MPQ has the three sides parallel to those of triangle ABC; therefore, they are similar.

Problem 1 of the Asian Pacific Mathematical Olympiad 1992

A triangle with sides a, b, and c is given. Denote by s the semi-perimeter, that is $s = (a + b + c) = 2$. Construct a triangle with sides $s - a$, $s - b$, and $s - c$. This process is repeated until a triangle can no longer be constructed with the side lengths given. For which original triangles can this process be repeated indefinitely?

Solution

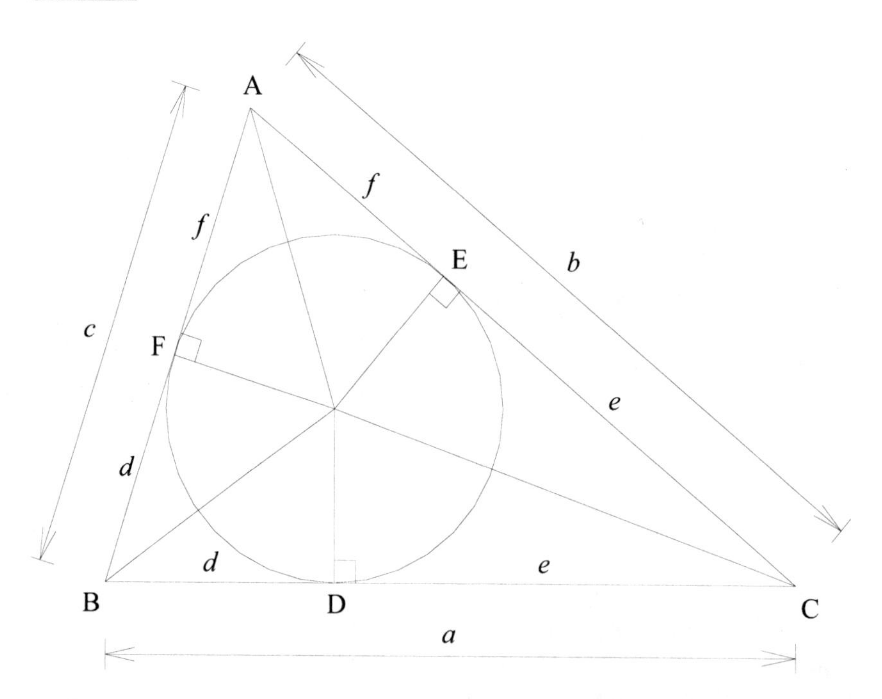

Draw the incircle of triangle ABC to tangent with the sides BC, AC and AB at D, E and F, respectively.

Let $BC = a$, $AC = b$, $AB = c$
$BD = BF = d$,
$CD = CE = e$
$AE = AF = f$

We have $s = (a + b + c)/2 = d + e + f = a + f = b + d = c + e$

So now we have $s - a = f$, $s - b = d$, and $s - c = e$

For the three sides to form a non-degenerate triangle, the sum of any two has to be greater than the third. So we must have

$f + d = c > e$, $f + e = b > d$, or $d + e = a > f$

For $c > e$, $\Rightarrow c > s - c \Rightarrow 2c > s \Rightarrow 4c > a + b + c \Rightarrow$
$3c > a + b$

Similarly, for $b > d \Rightarrow 3a > b + c$, and $a > f \Rightarrow$
$3b > a + c$

If one of those conditions is met, the process can be repeated, and the triangle can be constructed.

To construct the triangle draw a segment with the length of distance e; the ends of this segment are the vertices of the triangle that is under construction. Then from each end draw the circles with radii of d and f. These two circles intercept at another vertex of the triangle.

If the original triangle is equilateral, it will meet those conditions indefinitely since for an equilateral triangle $a = b = c$ and $d = e = f$ making the subsequent triangle also equilateral and the process keeps repeating indefinitely.

Problem 1 of Asian Pacific Mathematical Olympiad 1993

Let *ABCD* be a quadrilateral such that all sides have equal length and angle *ABC* is 60°. Let *l* be a line passing through *D* and not intersecting the quadrilateral (except at *D*). Let *E* and *F* be the points of intersection of *l* with *AB* and *BC* respectively. Let *M* be the point of intersection of *CE* and *AF*. Prove that $CA^2 = CM \times CE$.

Solution

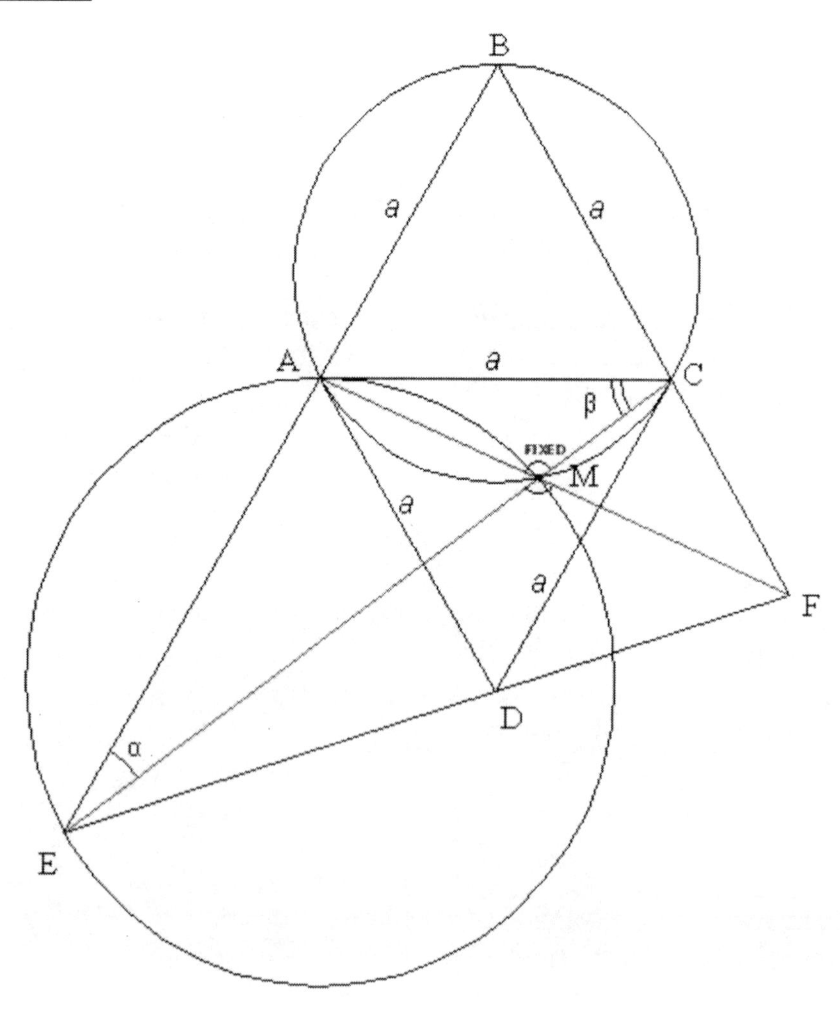

Let a be the length of the equilateral triangle ABC and ACD as shown, and let $\angle AEC = \alpha$ and $\angle ACE = \beta$

We have
$\alpha + \beta = 180° - \angle EAC = 180° - 120° = 60°$ (i)

We also have AE || CD and AD || CF; therefore, the two triangles EAD and DCF are similar, which causes
EA/a = a/CF.This makes the two triangles EBC and BCF to also be similar, and we have $\alpha = \angle CAF$.

Therefore, from (i) $\angle CAM + \beta = 60°$, and $\angle AMC = 120°$.

From there the two triangles EAC and AMC are similar because their respective angles equal. Therefore,

CE/CA = CA/CM, or $CA^2 = CM \times CE$.

Further observation

Let *ABCD* be a quadrilateral such that all sides have equal length and angle *ABC* is 60°. Let *l* be a line passing through *D* and not intersecting the quadrilateral (except at *D*). Let *E* and *F* be the points of intersection of *l* with *AB* and *BC* respectively. Let *M* be the point of intersection of *CE* and *AF*. Find the locus of point M.

Problem 1 of the Asian Pacific Mathematical Olympiad 2010

Let ABC be a triangle with $\angle BAC \neq 90°$. Let O be the circumcenter of the triangle ABC and let Γ be the circumcircle of the triangle BOC. Suppose that Γ intersects the line segment AB at P different from B, and the line segment AC at Q different from C. Let ON be a diameter of the circle Γ. Prove that the quadrilateral APNQ is a parallelogram.

Solution

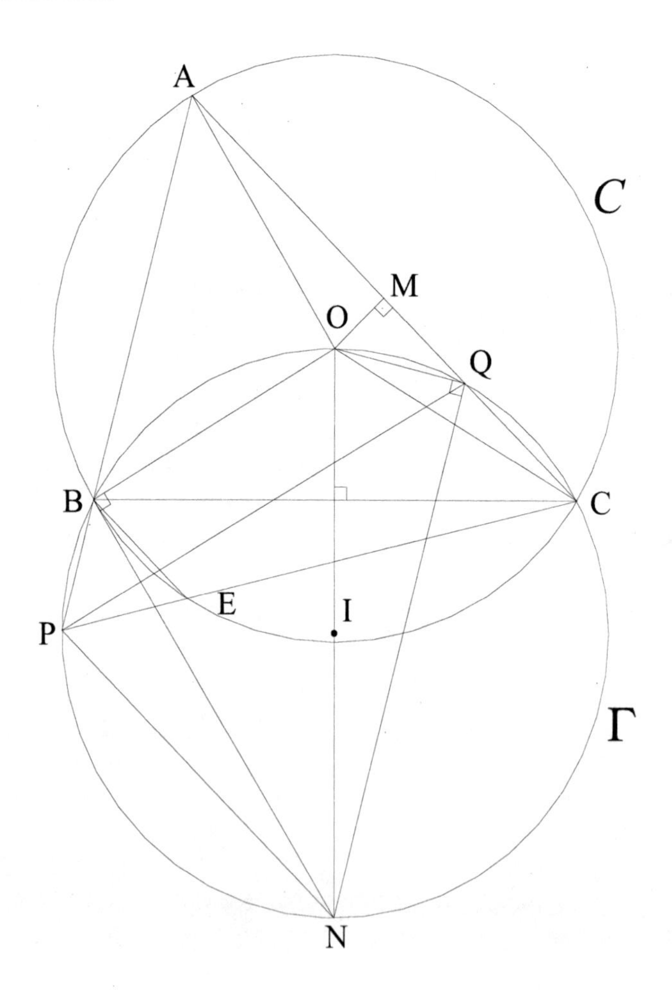

Let the circumcircle of triangle ABC be *C*, M be the midpoint of AC and E the intersection of *C* with PC. Also let r and R be the radii of *C* and Γ, respectively.

Consider two right triangles MOC and QON with $\angle$ONQ = $\angle$OCQ (subtends arc OQ). So they are similar; therefore,

MC/OC = QN/ON or MC/r = QN/2R, or

AC/QN = r/R, or $\angle$ABC = $\angle$NPQ.

We also have
$\angle$PQN = $\angle$PBN (subtends PN) = 180° − $\angle$OBN − $\angle$ABO
= 90° − $\angle$ABO

But $\angle$ABO = $\angle$BAO, $\angle$OBC = $\angle$OCB and $\angle$OAC = $\angle$OCA, or

$\angle$OCB + $\angle$OCA = $\angle$ACB = 90° − $\angle$ABO = $\angle$PQN

Now $\angle$BAC = 180° − $\angle$ABC − $\angle$ACB = 180° − $\angle$NPQ − $\angle$PQN = $\angle$PNQ

We also have $\angle$PNQ = $\angle$PCQ = $\angle$BAC or AP = PC and BE || AC and since BN is tangent to *C*, $\angle$NBE = $\angle$BCE = $\angle$BNP (subtends BP) or BE || PN.

Along with BE || AC, we have PN || AC.

Now combining with $\angle$BAC = $\angle$PNQ, we conclude that APNQ is a parallelogram.

Problem 1 of the Belarusian Mathematical Olympiad 2000

Find all pairs of integers (x, y) satisfying the equality
$y(x^2 + 36) + x(y^2 - 36) + y^2(y - 12) = 0$.

Solution

Rewrite the equation as follows

$$y^2 - 12y + 36 + x^2 + xy = \frac{36x}{y} \qquad \text{(i)}$$

The sum on the left is an integer, so the division on the right also must be an integer. Therefore, y must take on the values of

$y = \pm 1, \pm 2, \pm 3, \pm 4, \pm 6, \pm 9, \pm 12, \pm 36, \pm 1x, \pm 2x, \pm 3x, \pm 4x, \pm 6x, \pm 9x, \pm 12x, \pm 36x$

Now re-write (i) as follows

$$yx^2 + (y + 6)(y - 6)x + y(y - 6)^2 = 0 \qquad \text{(ii)}$$

Solving for x, $$x_{1\&2} = \frac{6 - y}{2y} \{ 6 + y \pm \sqrt{-3[(y - 2)^2 - 4^2]} \} \qquad \text{(iii)}$$

The first requirement is for the item under the square root to be non-negative which requires $|y - 2| \leq 4$, and the above list of the possible y values reduces to

$y = \pm 1, \pm 2, 3, 4, 6, \pm 1x, \pm 2x, \pm 3x, \pm 4x, \pm 6x, \pm 9x, \pm 12x, \pm 36x$

Now let $y = nx$ where n is an integer, equation (ii) can now be written as $(n^3 + n^2 + n)x^2 - 12n^2x + 36n - 36 = 0$. Solving for x, we have

$$x_{1\&2} = \frac{6}{n^3 + n^2 + n} [n^2 \pm \sqrt{n}] \qquad \text{(iv)}$$

The list is now reduced to $y = \pm 1, \pm 2, 3, 4, 6$ and $n = 1, 4, 9, 36$
Substituting the values $y = \pm 1, \pm 2, 3, 4, 6$ into equation (iii), and the values $y = x, 4x, 9x, 36x$ (iv) we have the following pairs of integers to satisfy the equality

(x, y) = (4, 4), (1, 4) and (0, 6)

Problem 1 of the British Mathematical Olympiad 2008

Find all solutions in non-negative integers a, b to $\sqrt{a} + \sqrt{b} = \sqrt{2009}$

Solution 1

Squaring both sides, we have $a + b + 2\sqrt{ab} = 2009$; now squaring them again, we have $(a + b - 2009)^2 = 4ab$, or
$a^2 - 2(b + 2009)a + b^2 + 2009^2 - 4018b = 0$
Solving for a, we obtain

$a = b + 2009 \pm \sqrt{2 \times 4018b} = b + 2009 \pm 14\sqrt{41\, b}$
Now for a to be an integer, 41b has to be the square of an integer, or $b = 41n^2$ where n is an integer.
Now $a = 41n^2 + 2009 \pm 14 \times 41n = 41n^2 + 2009 \pm 574n$
Note that a or b can not exceed 2009 and must not be negative, we have the following solutions when n = 1, 2, 3, 4, 5, 6 and 7
(b, a) = (41, 41x36), (41x4, 41x25), (41x9, 41x16), (41x16, 41x9), (41x25, 41x4), (41x36, 41), and (41x49, 0)

and since $\sqrt{a}$ and $\sqrt{b}$ are commutative; another series of solutions are
(a, b) = (41, 41x36), (41x4, 41x25), (41x9, 41x16), (41x16, 41x9), (41x25, 41x4), (41x36, 41), and (41x49, 0)

Solution 2

Let's write $\sqrt{a} + \sqrt{b} = \sqrt{2009}$ as $\sqrt{a} + \sqrt{b} = 7\sqrt{41}$

$\sqrt{a}$ takes on the values 0, $\sqrt{41}$, $2\sqrt{41}$, $3\sqrt{41}$, $4\sqrt{41}$, $5\sqrt{41}$,

$6\sqrt{41}$ and $7\sqrt{41}$ whereas $\sqrt{b}$ takes on the corresponding values

$7\sqrt{41}$, $6\sqrt{41}$, $5\sqrt{41}$, $4\sqrt{41}$, $3\sqrt{41}$, $2\sqrt{41}$, $\sqrt{41}$ and 0.

From there the same results as above are drawn.

Problem 1 of the Canadian Mathematical Olympiad 1969

Show that if $a_1/b_1 = a_2/b_2 = a_3/b_3$ and p_1, p_2, p_3 are not all zero, then

$(a_1/b_1)n = (p_1a_1n + p_2a_2n + p_3a_3n)/(p_1b_1n + p_2b_2n + p_3b_3n)$
for every positive integer n.

Solution

Adding a_1/b_1 ratio to the left onto the already existing equal ratios

$a_1/b_1 = a_2/b_2 = a_3/b_3$, we have

$a_1/b_1 = a_1/b_1 = a_2/b_2 = a_3/b_3$

Now raise to the n power for all

$(a_1/b_1)n = (a_1/b_1)n = (a_2/b_2)n = (a_3/b_3)n$

Now multiply both sides of different ratios with equal numbers p's

$(a_1/b_1)n = p_1a_1n /(p_1b_1n) = p_2a_2n /(p_2b_2n) = p_3a_3n /(p_3b_3n) =$

$(p_1a_1n + p_2a_2n + p_3a_3n)/ (p_1b_1n + p_2b_2n + p_3b_3n)$

Problem 1 of the Canadian Mathematical Olympiad 1971

DEB is a chord of a circle such that DE = 3 and EB = 5. Let O be the center of the circle. Join OE and extend OE to cut the circle at C. Given EC = 1, find the radius of the circle.

Solution

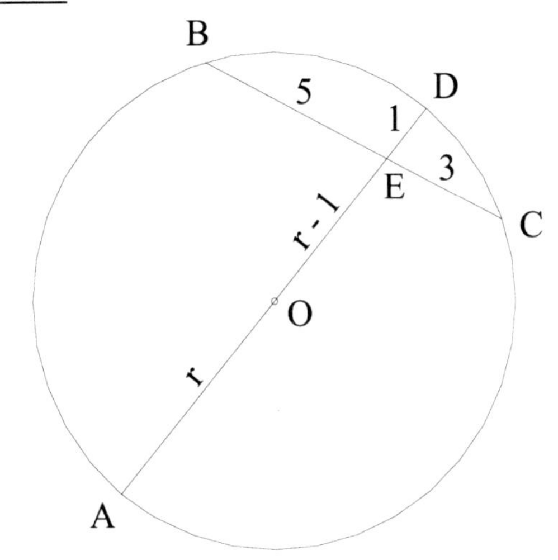

Extend CO to intercept the circle at A. Let r be the radius of the circle; we have

$OE = r - 1, \; OA = r$

Since $BE \times ED = EC \times EA$, we have $15 = 2r - 1$, or $r = 8$.

Problem 1 of the Canadian Mathematical Olympiad 1972

Given three distinct unit circles, each of which is tangent to the other two, find the radii of the circles which are tangent to all three circles.

Solution

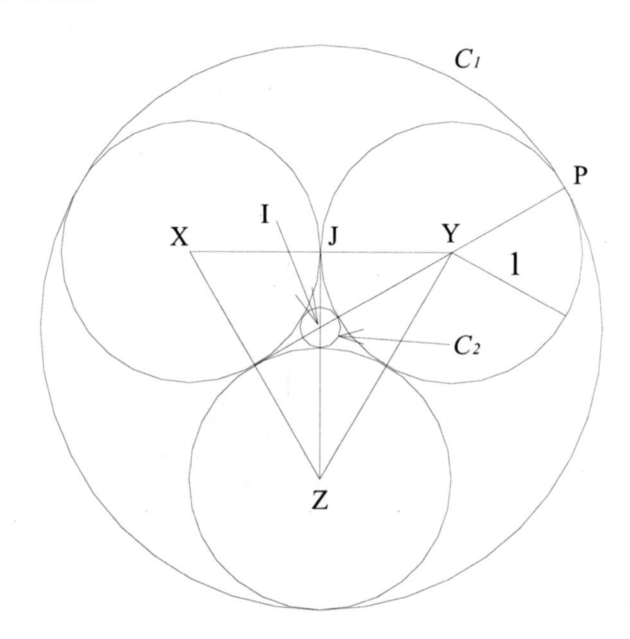

The three distinct unit circles are congruent with radii equal to 1. It's easily seen that the three centers X, Y and Z make an equilateral triangle since its lengths XY = YZ = ZX = 1.
The incircle/ centroid/ circumcenter I of this triangle will be the centers of the two circles which tangent to all three circles. For the larger circle C1 that tangents all three circle, its radius is

R = IP = YP + IY = 1 + IY. But IY = IZ = 2/3 altitude of triangle

XYZ = 2/3 ZJ = $2\sqrt{ZY^2 - JY^2}/3 = 2/\sqrt{3}$ so R = $1 + 2/\sqrt{3}$.
For the small C2 that tangents all three circle, its radius is r = R −

2YP = $1 + 2/\sqrt{3} - 2 = 2/\sqrt{3} - 1$.

Problem 1 of the Canadian Mathematical Olympiad 1975

Simplify

$$\sqrt[3]{\frac{1.2.4 + 2.4.8 + \ldots + n.\ 2n.\ 4n}{1.3.9 + 2.6.18 + \ldots + n.\ 3n.\ 9n}}$$

Solution

We have $1.2.4 + 2.4.8 + \ldots + n.2n.4n = 2.4\,(1^3 + 2^3 + 3^3 + \ldots + n^3)$

and $1.3.9 + 2.6.18 + \ldots + n.3n.9n = 3.9\,(1^3 + 2^3 + 3^3 + \ldots + n^3)$

and the ratio becomes $2.4 / (3.9) = 2^3/3^3$, or

$$\sqrt[3]{\frac{1.2.4 + 2.4.8 + \ldots + n.\ 2n.\ 4n}{1.3.9 + 2.6.18 + \ldots + n.\ 3n.\ 9n}} = 2/3$$

Problem 1 of the Canadian Mathematical Olympiad 1978

Let n be an integer. If the tens digit of n^2 is 7, what is the units digit of n^2?

Solution

Let the two least significant digits of number n be p and q with q being the least significant digit. The number n = XX.....XXpq (the X's represent the 'don't care' digits). Let's also denote U(k) as the units digit of number k.

We multiply the two numbers n together, and to easily understand, let's express the calculation in an ordinary way as follow:

```
     XX.....XXpq
  ×  XX.....XXpq   (small × denotes multiply)
     ...........ab
  +  ..........cd
     ..........feb
```

Note that b = U(q^2). If $q^2 > 9$, let q^2 = gb where g is the tens digit.

We then have
a = U(pq + g), and d = U(pq). Looking at the expression above,

e = U(a + d) is an even number plus g and is equal to 7 as is required by the problem.

Since the digit e = U(2pq + g) and is odd only when g is odd; therefore, q can not be equal to

0 (when $q^2 = 0$ and $g = 0$),
1 (when $q^2 = 1$ and $g = 0$),
2 (when $q^2 = 4$ and $g = 0$),
3 (when $q^2 = 9$ and $g = 0$),
5 (when $q^2 = 25$ and $g = 2$),

7 (when $q^2 = 49$ and $g = 4$),
8 (when $q^2 = 64$ and $g = 6$), or
9 (when $q^2 = 81$ and $g = 8$)

Let $q = 4$, $q^2 = 16$, $b = 6$, $g = 1$, $e = 7 = U(2pq + 1)$ so $U(2pq) = 6$, or

$U(pq) = 3$ but $q = 4$ and this is impossible.

Let $q = 6$, $q^2 = 36$, $b = 6$, $g = 3$, $e = 7 = U(2pq + 3)$ so $U(2pq) = 4$, or

$U(pq) = 2$ and $U(6p) = 2$ and $p = 2$ or $p = 7$.

So the units digit is 6; the tens digits is either 2 or 7.

Problem 1 of the Canadian Mathematical Olympiad 1980

If a679b is a five digit number (in base 10) which is divisible by 72, determine a and b.

Solution

The problem can be solved by trying different values for a from 1 to 9 to come up with the solution of $a = 3$, $b = 2$ and $a679b = 36792$.

A more savvy mathematical approach is following:

We have $a679b = 10000a + 6790 + b$ so $a679b / 72 = 10000a / 72 + 6790 / 72 + b / 72$

We know that if a679b is divisible by 72, the sum of the remainders of $10000a / 72$, $6790 / 72$ and $b / 72$ has to be divisible by 72. Let's denote $r(x / y)$ as remainder of x /y.

We have

$r (10000a/ 72) + r(6790/ 72) + r (b/ 72) = 72n$ (n is an integer), but $b < 10 < 72$, $r(b / 72) = b$,

and we have $r (10000a / 72) + 22 + b = 72n$

So we have to find a to satisfy the above condition. Still we substitute value for a:

For $a = 1$, $r (10000 / 72) = 64$ and $64 + 22 + b = 86 + b$ which is in range $[86, 95]$ and $\neq 72 n$

For $a = 2$, $r (20000 / 72) = 56$ and $56 + 22 + b = 78 + b$ which is in range $[78, 87]$ and $\neq 72 n$

For $a = 3$, $r (30000 / 72) = 48$ and $48 + 22 + b = 70 + b$ which is in range $[70, 79]$ and $= 72$ when $b = 2$.

So one solution is $a = 3$ and $b = 2$.

For $a = 4$, $r(40000 / 72) = 40$ and $40 + 22 + b = 62 + b$ which is in range $[62, 71]$ and $\neq 72n$

For $a = 5$, $r(50000 / 72) = 32$ and $32 + 22 + b = 54 + b$ which is in range $[54, 63]$ and $\neq 72n$

For $a = 6$, $r(60000 / 72) = 24$ and $24 + 22 + b = 46 + b$ which is in range $[46, 55]$ and $\neq 72n$

For $a = 7$, $r(70000 / 72) = 16$ and $16 + 22 + b = 38 + b$ which is in range $[38, 47]$ and $\neq 72n$

For $a = 8$, $r(80000 / 72) = 8$ and $8 + 22 + b = 30 + b$ which is in range $[30, 39]$ and $\neq 72n$

For $a = 9$, $r(90000 / 72) = 0$ and $0 + 22 + b = 22 + b$ which is in range $[22, 31]$ and $\neq 72n$

So $a = 3$ and $b = 2$ is the only solution.

Problem 1 of the Canadian Mathematical Olympiad 1981

For any real number t, denote by [t] the greatest integer which is less than or equal to t. For example: [8] = 8, [pi] = 3 and [−5/ 2] = -3. Show that the equation

$$[x] + [2x] + [4x] + [8x] + [16x] + [32x] = 12345$$

has no real solution.

Solution

Let $x = i + f$ where i is the integer part or integral part and f the fractional part of x. We have $f < 1$, and

$$[x] + [2x] + [4x] + [8x] + [16x] + [32x] = 63i + [f] + [2f] + [4f] + [8f] + [16f] + [32f]$$

Since $f < 1$, $[f] = 0$, and we have

$$63i + [f] + [2f] + [4f] + [8f] + [16f] + [32f] = 63i + [2f] + [4f] + [8f] + [16f] + [32f] = 12345 = 63 \times 195 + 60$$

So, $i = 195$, and

$$[2f] + [4f] + [8f] + [16f] + [32f] = 60 \qquad \text{(i)}$$

Since $\max[nf] = n - 1$, the maximum value of $[2f] + [4f] + [8f] + [16f] + [32f] = 1 + 3 + 7 + 15 + 31 = 57$.
Equation (i) is not possible.

Therefore, there is no f that can satisfy the equation in the problem, and thus there is no x.

Further observation

One can change the number 12345 to 12344 or 12343 and the problem is still valid.

Problem 1 of Canadian Mathematical Olympiad 1982

In the diagram, OBi is parallel and equal in length to AiAi+1 for i = 1; 2; 3 and 4 (A5 = A1). Show that the area of B1B2B3B4 is twice that of A1A2A3A4.

Solution

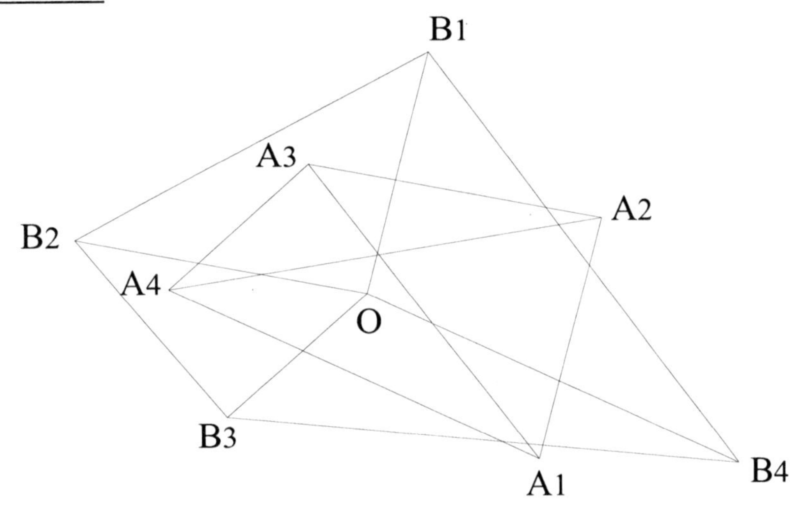

Let (Ω) denote the area of shape Ω. If we move the triangle OB1B2 with B2 –> A3 and O –> A2, (OB1B2) = (A1A2A3) since they have the equaled base A1A2 = OB1 and the same altitude from A3 (or B2 after the move).
We will see the same effect if we move triangle OB3B4 (B4 –> A1 and O –> A4), (OB3B4) = (A1A3A4)
Now adding the two areas
(A1A2A3) + (A1A3A4) = (OB1B2) + (OB3B4) (i)

Now move the triangle A1A2A4 (A1 –> O and A2 –>B1), (A1A2A4) = (OB1B4), and move the triangle A2A3A4 (A3 –> O and A4 –> B3), (A2A3A4) = (OB2B3)
Adding the previous two areas
(A1A2A4) + (A2A3A4) = (OB1B4) + (OB2B3) (ii)

Now adding the sides of (i) and (ii)

2 (A1A2A3A4) = (B1B2B3B4)

Problem 1 of Canadian Mathematical Olympiad 1983

Find all positive integers w, x, y and z which satisfy $w! = x! + y! + z!$

Solution

We have
$w! = z!\,(z+1)(z+2)\ldots(w-1)\,w$
$x! = z!\,(z+1)(z+2)\ldots(x-1)\,x$
$y! = z!\,(z+1)(z+2)\ldots(y-1)\,y$

We know that $w > x$, $w > y$ and $w > z$. These are possible combinational conditions for x, y and z

1) $x > y > z$
2) $x > y = z$
3) $x = y > z$
4) $x = y = z$

Since x, y and z are interchangeable, those cases do cover all the scenarios. Write the original equation $w! = x! + y! + z!$ as
$z!(z+1)(z+2)\ldots(w-1)w = z!(z+1)(z+2)\ldots(x-1)x + z!(z+1)(z+2)\ldots(y-1)y + z!$ (i)

Case 1) For $x > y > z$
i) If $x > y > z + 1$ (or $x > y \geq z + 2$), equation (i) can be written as
$(z+1)(z+2)\ldots(w-1)\,w = (z+1)(z+2)\ldots(x-1)\,x + (z+1)(z+2)\ldots(y-1)\,y + 1$

The two series on the right sides are both even numbers since each is a product of at least two consecutive numbers. Adding 1 and it becomes an odd number whereas the left side is an even number for the same reason. So this case is impossible.

ii) If $x > y = z + 1$, we have
$(z+1)(z+2)\ldots(w-1)\,w = (z+1)(z+2)\ldots(x-1)\,x + y + 1$

The series on the right side is an even number. Adding 1 and it becomes an odd number. Therefore, y can not be an even number since the series on the left is an even number.

Now for y as an odd number, we have

$(z+1)(z+2)\ldots(w-1)w - (z+1)(z+2)\ldots(x-1)x = y+1$
but since $w > x$, we have to prove this equality to be true
$(z+1)(z+2)\ldots(x-1)x\ [(x+1)(x+2)\ldots(w-1)w-1] = y+1$ (ii)

But since $x > 1$, $2x > x+1$ and product

$[z+1)(z+2)\ldots(x-1)[(x+1)(x+2)\ldots(w-1)w-1] > 2$ or
$[z+1)(z+2)\ldots(x-1)[(x+1)(x+2)\ldots(w-1)w-1]\ x > 2x > x+1 > y+1$. Therefore, (ii) is also impossible.

<u>Case 2)</u> For $x > y = z$, the original equation (i) can now be written

$$z!(z+1)(z+2)\ldots(w-1)w = z!(z+1)(z+2)\ldots(x-1)x + 2z!$$
or

$$(z+1)(z+2)\ldots(w-1)w = (z+1)(z+2)\ldots(x-1)x + 2 \quad \text{(iii)}$$

Now if $x > y+1 = z+1$, the series on the right is an even number. Adding 2 to it and the right side is an even number. Since $w > x$, we can write (iii) as

$$(z+1)(z+2)\ldots(x-1)x\ [(x+1)(x+2)\ldots(w-1)w-1] = 2$$

This is impossible since the series on the left has more than one multiplier > 2.

Now if $x = y+1 = z+1$, we write (iii) as

$(z+1)(z+2)\ldots(w-1)w = (z+1)(z+2)\ldots(x-1)x + 2(z+1)(z+2)\ldots(x-1)$

Dividing both side by $(z+1)(z+2)\ldots(x-1)$, we have

$$x(x+1)(x+2)\ldots(w-1)w = x+2 \qquad \text{(iv)}$$

but since $w > x = y + 1 = z + 1$ the minimum value of w is 3 or $w > 2$, or

$2w > w + 2$ and $x(x+1)(x+2)\ldots(w-1)w > 2w$ since the left side has more than one multiplier > 2, or

$x(x+1)(x+2)\ldots(w-1)w > 2w > w+2 > x+2$ and (iv) is also impossible.

<u>Case 3)</u> If $x = y > z$, we have

$(z+1)(z+2)\ldots(w-1)w = 2(z+1)(z+2)\ldots(x-1)x + 1$

The number on the right is odd, and the one on the left is even. This case is also impossible.

<u>Case 4)</u> If $x = y = z$, we have

$w! = 3x!$ which only occurs when $1.2.3 = 3(1.2)$ or $w = 3$ and $x = y = z$

Problem 1 of Canadian Mathematical Olympiad 1984

Prove that the sum of the squares of 1984 consecutive positive integers cannot be the square of an integer.

Solution

Let's write the sum as follows $S = n^2 + (n + 1)^2 + (n + 2)^2 + \ldots + (n + 1983)^2$

Now expanding the squares, we have
$S = 1984n^2 + 2n\,(1 + 2 + 3 + \ldots + 1983) + 1^2 + 2^2 + 3^2 + \ldots + 1983^2$

According to the Faulhaber's formulas

$1 + 2 + 3 + \ldots + n = n\,(n + 1)/2$, and
$1^2 + 2^2 + 3^2 + \ldots + n^2 = n(n + 1)(2n + 1)/6$

S now becomes $S = 1984n^2 + 1983 \times 1984n + 1983 \times 1984 \times 3967/\,6 = 992(\,2n^2 + 1983 \times 2n + 661 \times 3967\,)$

For this to be the square of an integer we must have
$2n^2 + 1983 \times 2n + 661 \times 3967 = 992m^2$ (i)

but we note that the product 661 x3967 is an odd number, and the sum on the left is an odd number whereas on the right $992m^2$ is an even number. Thus we can not find an integer m to satisfy (i).

Therefore, the sum of the squares of 1984 consecutive positive integers cannot be the square of an integer.

Further observation

This problem works for year 2010 even though the solution is different and much more difficult.

Problem 1 of the Canadian Mathematical Olympiad 1985

The lengths of the sides of a triangle are 6, 8 and 10 units. Prove that there is exactly one straight line which simultaneously bisects the area and perimeter of the triangle.

Solution

We note that ABC is a right triangle since $6^2 + 8^2 = 100 = 10^2$. Let AB = 6, AC = 8 and BC = 10, and let the straight line in question be named *l*. We know that *l* can not pass through any of the vertices since when passing through a vertex, it has to bisect the opposing side of the triangle in order to bisect its area, and therefore, it will not bisect the perimeter.
So there are three distinct scenarios for the positions of line *l* as follow:

Case 1

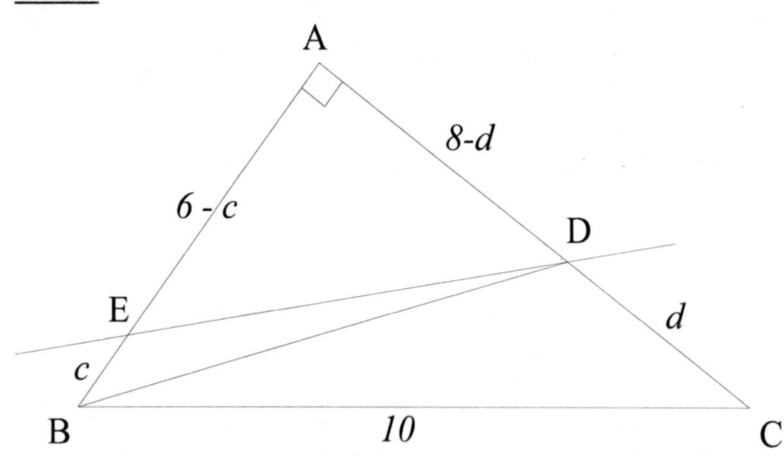

When *l* cuts through and intercepts AB and AC at E and D, respectively as shown on the graph. Let BE = c and DC = d, then EA = 6 –c and AD = 8 –d. We have $c + d = 2$, and one half of the area is $6d + c(8 - d) = 24$. Substituting c from above we have

$d^2 - 14d + 72 = 0$, or $d = 7 \pm\sqrt{-23}$ and are invalid since $-23 < 0$.

Case 2

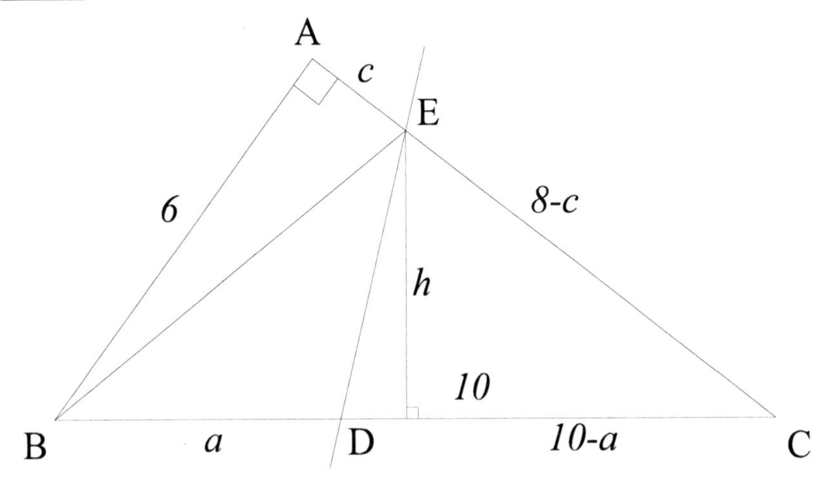

When l cuts through and intercepts AC and BC at E and D, respectively as shown on the graph. Let AE = c and BD = a, then EC = 8 –c and DC = 10 –a. From E draw the altitude h to BC.

We have
$a + c + AB = \frac{1}{2}$ perimeter $= 12 - 6 = 6$, or $\quad c = 6 - a$

Since we assume line l to bisect the area, the two halves of area are equal

$6c + ah = h(10 - a)$, or $h(a - 5) = -3c$, or $\quad h = 3c/(5 - a)$

and each half of the area of the triangle is now
$h(10 - a) = 6 \times 8/2 = 24$, or

$3c(10 - a)/(5 - a) = 24$, or $\quad 10c - ac = 40 - 8a \qquad$ (i)

Now substituting $c = 6 - a$ to (i), we have $\quad a^2 - 8a + 20 = 0$

which is supposed to have two roots as $a = 4 \pm \sqrt{-4}$ which are invalid since $-4 < 0$.

Case 3

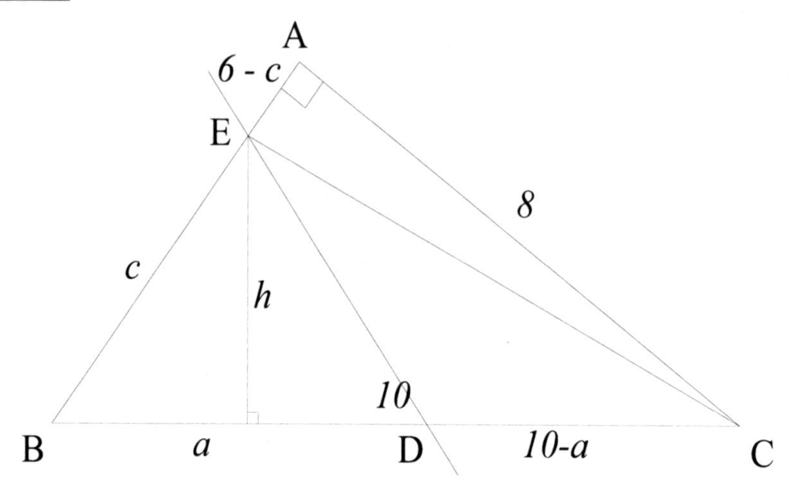

When *l* cuts through and intercepts AB and BC at E and D, respectively as shown on the graph. Let BE = c and BD = a, then EA = 6 –c and DC = 10 –a. From E draw the altitude h to BC.

We have $a + c = 12$.

And each half of the area is $8(6 - c) + h(10 - a) = ah$, or $h = (4c - 24)/(5 - a)$, but in this case $ah = 24$ or $h = 24/a$, and the above equation becomes

$(4c - 24)/(5 - a) = 24/a$, or $ac = 30$ now combining with $a + c = 12$, we come up with the quadratic equation $a^2 - 12a + 30 = 0$ which has the two roots as

$a = 6 \pm \sqrt{6}$ which are valid since $6 > 0$ and from there $c = 6 \mp \sqrt{6}$, but note that c must be smaller than AB = 6, so the root $c = 6 + \sqrt{6}$ is invalid and so is $a = 6 - \sqrt{6}$. The only valid solution is now $a = a = 6 + \sqrt{6}$ and $c = 6 - \sqrt{6}$.

Problem 1 of the Canadian Mathematical Olympiad 1986

In the diagram line segments AB and CD are of length 1 while angles ABC and CBD are 90° and 30° respectively. Find AC.

Solution

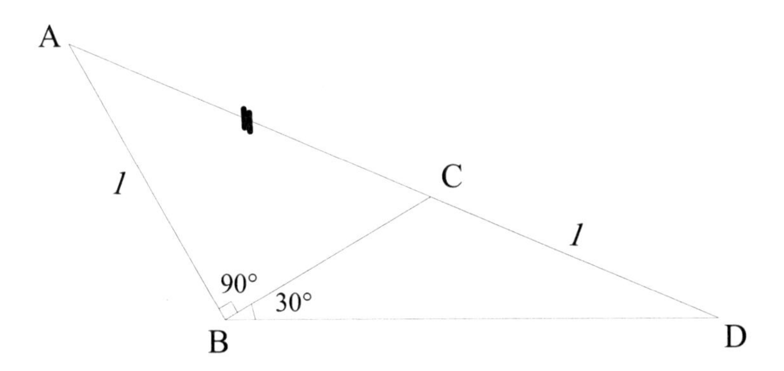

We have $AC^2 = 1 + BC^2$ and $BC/\sin\angle D = CD/\sin 30° = 2$, or $\sin\angle D = \frac{1}{2} BC$.

We also have $(AC + 1)/\sin 120° = AB/\sin\angle D = 1/\sin\angle D = 2/BC$.

or $(AC + 1)/\sqrt{3} = 1/\sqrt{AC^2 - 1}$ or $(AC + 1)^3 (AC - 1) = 3$ or

$AC^4 + 2AC^3 - 2AC + AC^2 - 4 = 0$, or $AC(AC^3 - 2) + 2(AC^3 - 2) = 0$, or

$$(AC^3 - 2)(AC + 2) = 0 \text{ but } AC + 2 > 0.$$

Therefore, $AC^3 - 2 = 0$, and $AC = \sqrt[3]{2}$.

Further observation

This problem is the same as problem 3 of the Irish Mathematical Olympiad 2010.

Problem 1 of the Canadian Mathematical Olympiad 1988

For what values of b do the equations: $1988x^2 + bx + 8891 = 0$ and $8891x^2 + bx + 1988 = 0$ have a common root?

Solution

The roots for the quadratic equation $1988x^2 + bx + 8891 = 0$ are

$$x_{1\&2} = (-b \pm \sqrt{b^2 - 4 \times 1988 \times 8891}\,) / (2 \times 1988)$$

The roots for the quadratic equation $8891x^2 + bx + 1988 = 0$ are

$$x_{3\&4} = (-b \pm \sqrt{b^2 - 4 \times 1988 \times 8891}\,) / (2 \times 8891)$$

For them to have the same root, equate the roots

For $x1 = x3$ or $x2 = x4$, we have

$b^2 = b^2 - 4 \times 1988 \times 8891$ which has no solution.

For $x1 = x4$ or $x2 = x3$, we have

$b^2 (10879^2 - 6903^2) = 10879^2 \times 4 \times 1988 \times 8891$ or $b = 10879$.

Solution is $b = 10879$.

Problem 1 of the Canadian Mathematical Olympiad 1992

Prove that the product of the first n natural numbers is divisible by the sum of the first n natural numbers if and only if n + 1 is not an odd prime.

Solution

The product of the first n natural numbers is
$1.2.3.4 \ldots\ldots\ldots (n-1)\, n$

The sum of the first n natural numbers is
$1 + 2 + 3 + 4 + \ldots + (n-1) + n = 0.5n\,(n+1)$

Let k be the resultant of the product divided by the sum, we have

$$k = \frac{1.2.3.4\ldots (n-1)\, n}{0.5n\,(n+1)} = \frac{2.2.3.4\ldots (n-1)}{n+1}$$

The following are possibilities for n + 1

a) n +1 is an even number. Let $n + 1 = 2k$ or $n - 1 = 2k - 2 = 2(k - 1)$ and

$$k = \frac{2.2.3.4\ldots (n-1)}{n+1} = \frac{2.2.3.4\ldots (n-2)\, 2(k-1)}{2k} =$$
$$= \frac{2.2.3.4\ldots(k-1)\, k\, (k+1)\ldots (n-3)\,(n-2)\,(k-1)}{k}$$
$$= 2.2.3.4\ldots(k-2)\,(k-1)\,(k+1)(k+2)\ldots (n-3)\,(n-2)\,(k-1)$$

So the first case of n not being an odd prime satisfies the problem.

b) n + 1 is an odd number:

i) It's a prime number

When n + 1 is a prime number it can not be factored out to smaller numbers, and thus the product is then not divisible by the sum.

$$k = \frac{2.2.3.4\ldots(n-1)}{n+1}$$

ii) It's not a prime number

When n + 1 is not a prime number it can be factored out to two or more numbers that are smaller than itself $n + 1 = m(m+l)(m+p)\ldots$ and $(m + p) < n - 1$ and thus the product is then divisible by the sum.

Note: n in the problem must be > 2.

Problem 1 of Canadian Mathematical Olympiad 1993

Determine a triangle for which the three sides and an altitude are four consecutive integers and for which this altitude partitions the triangle into two right triangles with integer sides. Show that there is only one such triangle.

Solution

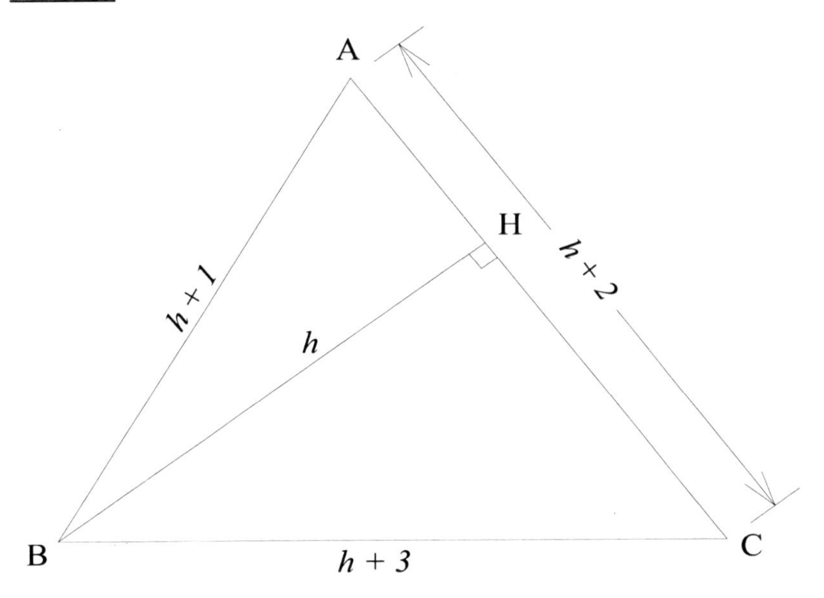

From B draw altitude BH to AC; let BH = h.

We then have $AB = h + 1$, $AC = h + 2$, $BC = h + 3$

$AH^2 = AB^2 - BH^2$ or $AH^2 = (h + 1)^2 - h^2$ or $AH = \sqrt{2h + 1}$
Similarly, $HC^2 = BC^2 - BH^2$ or $HC^2 = (h + 3)^2 - h^2$,

or $HC = \sqrt{6h + 9}$

$$AC = AH + HC \quad \text{or} \quad h + 2 = \sqrt{2h + 1} + \sqrt{6h + 9} \qquad \text{(i)}$$

Now square both sides of (i), we have

$$h^2 + 4h + 4 = 2h + 1 + 6h + 9 + 2\sqrt{(2h + 1)(6h + 9)}, \text{ or}$$

$$(h^2 - 4h - 6)^2 = 4(12h^2 + 24h + 9),$$

or $$h[\,h(h - 8) - 44\,] = 48$$

From here we conclude that h and $h(h - 8) - 44$ can take on these values (1, 48), (2, 24), (3, 16), (4, 12), (6, 8), (8, 6), (12, 4), (16, 3), (24, 2), (48, 1).

Further, $h > 8$ otherwise $h - 8 < 0$ and the left side is negative. We came up with $h = 12$. From there $AB = 13$, $AC = 14$ and $BC = 15$.

To show that there is only one such triangle, we tried drawing the altitudes from other vertices to their bases, and came up with equations that have no solutions. For example, from A drawing the altitude to BC, using the same procedure, we came up with the equation

$h(h^2 - 24) = 48$, and there is no integer h to satisfy it.

Problem 1 of China Mathematical Olympiad 2010

Circle $\Gamma 1$ and $\Gamma 2$ intersect at two points A and B. A line through B intersects $\Gamma 1$ and $\Gamma 2$ at points C and D, respectively. Another line through B intersects $\Gamma 1$ and $\Gamma 2$ at points E and F, respectively. Line CF intersects $\Gamma 1$ and $\Gamma 2$ at points P and Q, respectively. Let M and N be the midpoints of arcs PB and QB, respectively. Prove that if CD = EF, then C, F, M and N are concyclic.

Solution

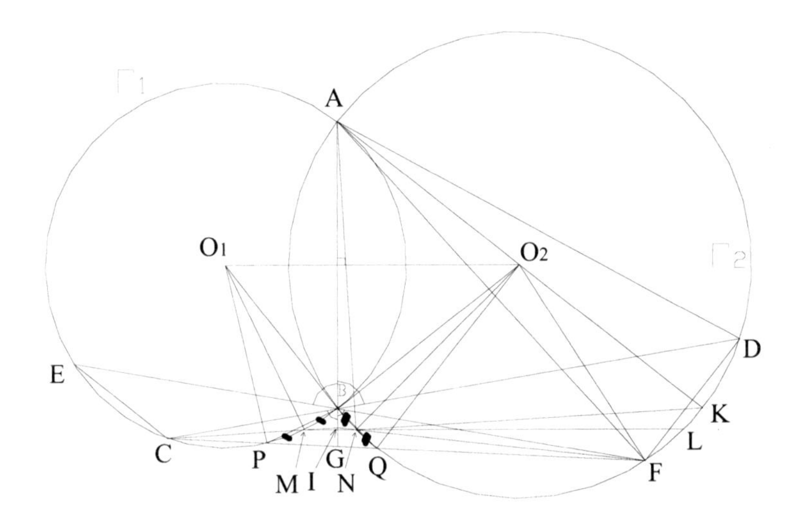

Extend AB to meet CF at G. We are going to prove that BG is the bisector of $\angle CBF$. We have $CB \times CD = CQ \times CF$, and

$FB \times EF = FP \times CF$.

Dividing the two above equations, knowing CD = EF,

we get $\dfrac{CB}{FB} = \dfrac{CQ}{FP}$.

We also have $PG \times CG = GB \times GA = QG \times FG$,

or $\dfrac{QG}{PG} = \dfrac{CG}{FG} = \dfrac{QG + CG}{PG + FG} = \dfrac{CQ}{PF}$

It follows that $\frac{CG}{FG} = \frac{CB}{FB}$, or $\angle CBG = \angle FBG$ and BG is the bisector of $\angle CBF$.

So now the three bisectors CM, FN and BG coincide. Let them meet at I on BG. We now have $IM \times IC = IB \times IA = IN \times IF$, or IM/IN = IF/IC, or the two triangles IMN and IFC are similar, meaning $\angle IMN = \angle IFC$, but

$\angle IMN + \angle CMN = 180°$, or $\angle CMN + \angle NFC = 180°$ and C, F, M and N are concyclic.

Further observation

Let's prove that MN || O*1*O*2* where O*1* and O*2* are centers of Γ*1* and Γ*2*, respectively. Since $\angle CBG = \angle FBG$, we have $\angle ABD = \angle FBG$, or $AD = AF$.

Let K be the midpoint of arc FD, AK is then the diameter of Γ*2*.

$\angle MCP = ½ \angle BCP = ½$ arc (DF – BQ) = arc (KF – NQ) or $\angle MCP + \angle NFQ$ = arc (FK). Extend MN to meet Γ*2* at L, $\angle LNF = \angle MCP$; therefore, $\angle KNL = \angle NFQ = \angle BAN$ (subtending arc NB = NQ). But $\angle ANK = 90°$ or $AN \perp NK$; therefore, $NL \perp AG$ or MN || O*1*O*2*.

Problem 1 of the Belarusian Mathematical Olympiad 2000

Find all pairs of integers (x, y) satisfying $3xy - x - 2y = 8$.

Solution

Let $x = y + z$ where z is an integer. The equation can be written as $3y^2 + 3(z - 1)y - z - 8 = 0$. Solving for y, we have

$$y_{1\&2} = \frac{1}{6}(3 - 3z \pm \sqrt{9z^2 - 6z + 105}) \qquad \text{(i)}$$

Now assume $9z^2 - 6z + 105$ is a square, the first requirement for y to be an integer.

Let $9z^2 - 6z + 105 = n^2$ where n is an integer, or $9z^2 - 6z + 105 - n^2 = 0$

Solving for z, we have $z_{1\&2} = \frac{1}{9}[3 \pm \sqrt{9(n^2 - 104)}]$

So now $n^2 - 104$ has to be a square. Let $n^2 - 104 = m^2$ (ii)

To satisfy the equation, both n and m have to be either odd or even.

a) For the case of both n and m being even integers, let $n = 2p$ and $m = 2q$, where p and q are both integers, and now (ii) becomes $4p^2 - 104 = 4q^2$, or $(p - q)(p + q) = 26$, and $p - q$ and $p + q$ can take on these values
$(p - q, p + q) = (1, 26), (2, 13), (13, 2)$ or $(26, 1)$

or $2p = 27$ or $2p = 15$ and neither of these is possible because 2p is an even integer.

b) If they are both odd integers, let $n = 2p + 1$ and $m = 2q + 1$, where p and q are both integers. Now (ii) becomes $4p^2 + 4p - 104 = 4q^2 + 4q$, or $(p - q)(p + q + 1) = 26$, and $p - q$ and $p + q + 1$ can take on these values

$(p - q, p + q + 1) = (1, 26), (2, 13), (13, 2)$ or $(26, 1)$, or $p = 13$,

$p = 7$ which make $n = 27$ or $n = 15$, respectively.

For $n = 27$, $z_{1\&2} = 78/9$ or -8 for which $z = -8$ is the only integer solution.

Substituting $z = -8$ into (i), we have $y = 0$ and $y = 9$, and subsequent substitution of these values of y into the equation of the problem, we have $y = 0$, $x = -8$ and $y = 9$, $x = 1$

For $n = 15$, $z_{1\&2} = 4$ or $-30/9$ for which $z = 4$ is the only integer solution.

Substituting $z = 4$ into (i), we have $y = -4$ and $y = 1$, and subsequent substitution of these values of y into the equation of the problem, we have $y = -4$, $x = 0$ and $y = 1$, $x = 5$

Answers: All pairs of integers (x, y) satisfying $3xy - x - 2y = 8$ are

$(x, y) = (-8, 0), (0, -4), (1, 9)$ and $(5, 1)$

Problem 1 of the Ibero-American Mathematical Olympiad 1988

The measures of the angles of a triangle is an arithmetic progression and its altitudes is also another arithmetic progression. Prove that the triangle is equilateral.

Solution

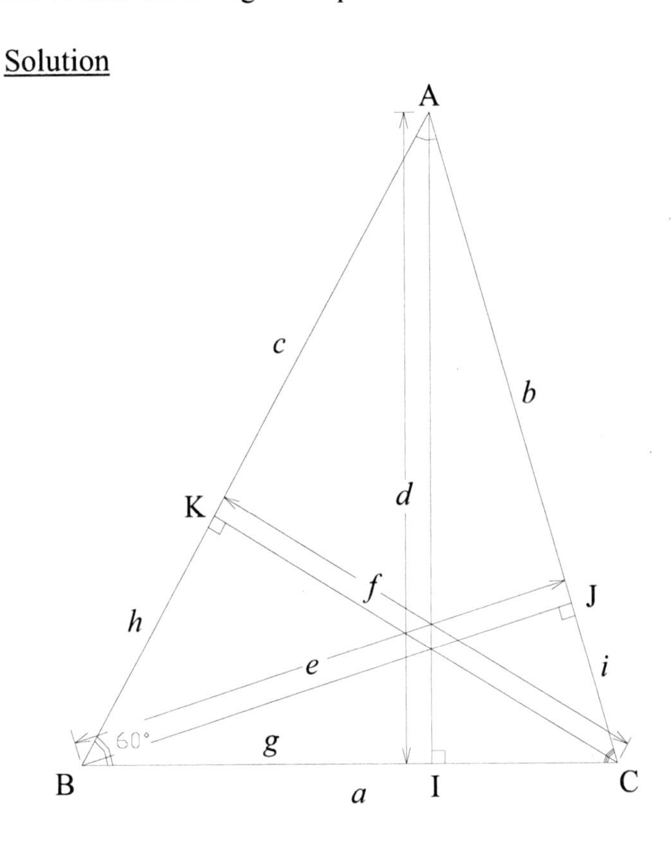

Let I, J, and K be the feet of A, B and C to BC, AC and AB, respectively.

Now let BC = a, AC = b, AB = c, AI = d, BJ = e, CK = f, BI = g, CJ = i and BK = h. Also let $\angle BAC = \alpha$, $\angle ABC = \beta$ and ACB = γ.

Assume α is the smallest angle of the triangle and ε is the angle of common difference.

We have $\beta = \alpha + \varepsilon$, $\gamma = \alpha + 2\varepsilon$, but the sum of the angles is 180°, we then have $3(\alpha + \varepsilon) = 180°$, or $\beta = \alpha + \varepsilon = 60°$, and $\alpha = 120° - \gamma$.

Now we need to prove a = c for the triangle ABC to be equilateral.

Since $\beta = 60°$, we have $a = 2h$, $c = 2g$ and $f^2 = a^2 - h^2 = 3h^2$ or $f = h\sqrt{3} = a\sqrt{3}/2$.

Similarly $d = c\sqrt{3}/2$ and since d, e, and f form another arithmetic progression, we have

$$e = (f + d)/2 = (a + c)\sqrt{3}/4 \qquad \text{(i)}$$

We also have $\sin\alpha = e/c$ and $\sin\gamma = e/a$ or

$$\sin\alpha = \sin(120° - \gamma) = \sqrt{3}/2 \cos\gamma + \tfrac{1}{2}\, e/a = e/c \qquad \text{(ii)}$$

but $\cos\gamma = i/a$, (ii) becomes $\sqrt{3}/2\; i/a + \tfrac{1}{2}\, e/a = e/c$ (iii)

Apply the Pythagorean's theorem to right triangle BJC, we have

$$i = \sqrt{a^2 - e^2}$$

Now substituting i and e from (i) to (iii), we have

$$a^4 + c^4 + a^3c + ac^3 - 4a^2c^2 = 0, \text{ or}$$

$$(a - c)^2 (a^2 + 3ac + c^2) = 0,$$

or $a = c$.

Problem 1 of the International Mathematical Olympiad 1998

In the convex quadrilateral ABCD, the diagonals AC and BD are perpendicular and the opposite sides AB and DC are not parallel. Suppose that the point P, where the perpendicular bisectors of AB and DC meet, is inside ABCD. Prove that ABCD is a cyclic quadrilateral if and only if the triangles ABP and CDP have equal areas.

Solution

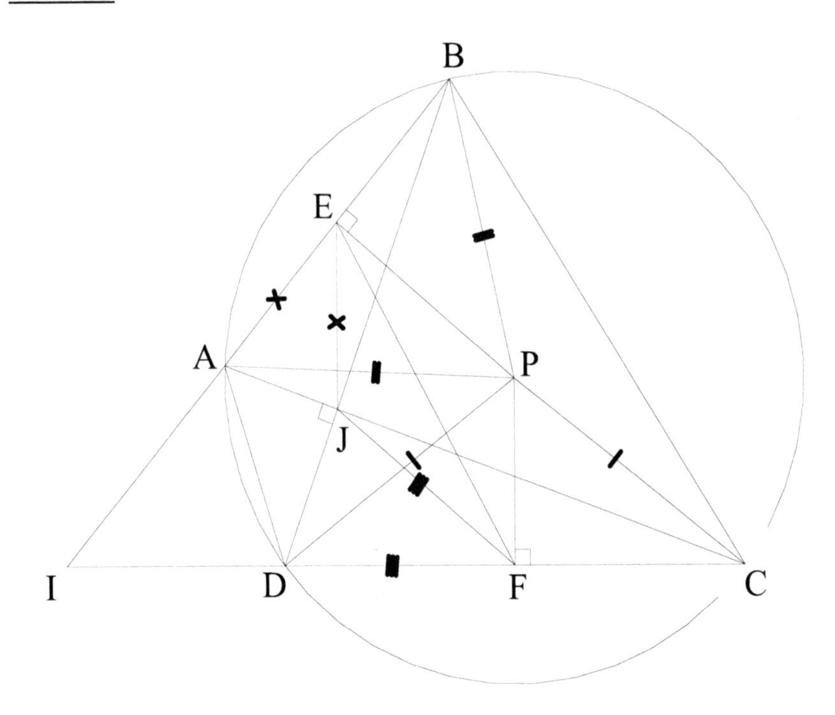

Let AC intercept BD at J, AB intercept CD at I, E and F be the midpoints of AB and CD, respectively. We have

$\angle$EAJ = $\angle$EJA, $\angle$EBJ = $\angle$EJB, $\angle$FDJ = $\angle$FJD, $\angle$FJC = $\angle$FCJ and JE = ½ AB, JF = ½ CD. Triangles *ABP* and *CDP* having equal areas gives us

PE × AB = PF × CD, or

PE/PF = CD/AB = JF/JE (i)

But $\angle$EJA + $\angle$EJB + $\angle$FJD + $\angle$FJC = 180°
Or $\angle$EJB + $\angle$FJC = 180° − $\angle$EJA − $\angle$FJD

Therefore,
$\angle$EJF = $\angle$EJB + 90° + $\angle$FJC = $\angle$EBJ + 90° + $\angle$FCJ = 180° − $\angle$EIF = $\angle$EPF

Combining d with (i) and the fact that they share segment EF, the triangles JEF and PFE are congruent, and EPFJ is a parallelogram.

It follows that PE = JF = DF and PF = JE = AE, and the two triangles AEP and PFD are congruent which causes PA = PD, or P is the center of the circumcircle passing through A, B, C and D. ABCD is then a cyclic quadrilateral.

Conversely, if ABCD is a cyclic quadrilateral, P is the center of the circumcircle. Since AC is perpendicular to BD, the sum of the angles subtending arcs AB plus CD equal 90°.

Therefore, $\angle$APB + $\angle$CPD = 180° or $\angle$APE + $\angle$FPD = 90° or $\angle$APE = $\angle$DPF and the two triangles APE and DPF are congruent (similar triangles with PA = PD). Therefore, triangles ABP and CDP with each having twice the areas of the triangles APE and DPF, respectively, have equal areas.

Problem 1 of Irish Mathematical Olympiad 1994

Let x, y be positive integers with $y > 3$ and $x^2 + y^4 = 2[(x-6)^2 + (y+1)^2]$.

Prove that $x^2 + y^4 = 1994$.

Solution

Expanding and combining the same terms, we have

$x^2 - 24x - y^4 + 2y^2 + 4y + 74 = 0$

Solving for x, we have

$$x_{1\&2} = 12 \pm \sqrt{y^4 - 2y^2 - 4y + 70} \qquad \text{(i)}$$

Therefore, $y^4 - 2y^2 - 4y + 70$ has to be a square of an integer and one possible scenario is that $2y^2 + 4y - 70 = 0$ for the term inside the square root to equal y^4, and the positive value for y is $y = 5$.

Now substituting $y = 5$ into (i), the positive value for x is $x = 37$.

Hence, $37^2 + 5^4 = 1994$.

Further observation

If y is not bounded by the condition $y > 3$, we have $y = 3$ which makes $x = 1$ or $x = 23$.

Problem 1 of Irish Mathematical Olympiad 2007

Let r, s and t be the roots of the cubic polynomial

$$p(x) = x^3 - 2007x + 2002$$

Determine the value of $\frac{r-1}{r+1} \cdot \frac{s-1}{s+1} \cdot \frac{t-1}{t+1}$

Solution

Expanding

$$\frac{r-1}{r+1} \cdot \frac{s-1}{s+1} \cdot \frac{t-1}{t+1} = \frac{3rst - 3 + rt + st + rs - s - r - t}{rst + rt + st + rs + s + r + t + 1} \qquad \text{(i)}$$

Since r, s and t are the roots, we can write p(x) as

$(x - r)(x - s)(x - t) = x^3 - (s + r + t)x^2 + (rt + st + rs)x + rst =$
$x^3 - 2007x + 2002$

or $s + r + t = 0$, $rt + st + rs = -2007$, and $rst = -2002$

Substituting them into (i), we have

$$\frac{r-1}{r+1} \cdot \frac{s-1}{s+1} \cdot \frac{t-1}{t+1} = \frac{-3\times2002 - 3 - 2007}{-2002 - 2007 + 1} = \frac{-8016}{-4008} = 2$$

Problem 1 of the British Mathematical Olympiad 2006

Triangle ABC has integer-length sides, and AC = 2007. The internal bisector of $\angle$BAC meets BC at D. Given that AB = CD, determine AB and BC.

Solution

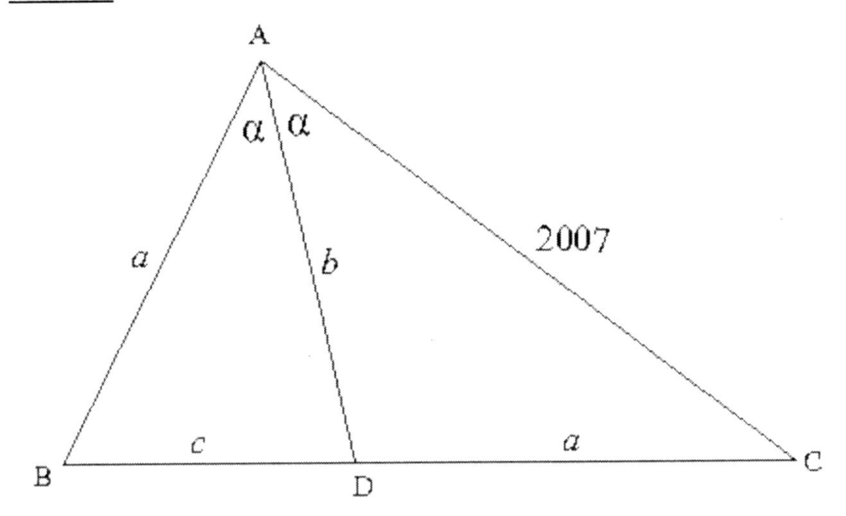

Let AB = CD = a, AD = b, BD = c and $\angle BAC = 2\alpha$. Since AD is the bisector of $\angle$BAC, we have BD/AB = DC/AC or $a^2 = 2007c = 3^2 \times 223c$.
For a to be an integer the product 223c has to be a square of an integer, or $c = 223n^2$ *where n is a positive integer, and* $a^2 = 669^2n^2$ *or* $a = 669n$.
For n = 1, a = 669, c = 223, and BC+AB = 2a+c = 1561 < AC which is impossible since the sum of the two sides of a triangle must be greater than the other side.

For n = 2, a = 669x2 = 1338, c = 892, a + c = 2230, and BC + AB = 2a + c = 3568 which is possible.
For $n \geq 3$, $c \geq 223 \times 9 = 2007$, $a + c \geq a + 2007$ or $BC = a + c \geq a + 2007 = AB + AC$ which is again impossible.

So the only solution is AB = 1338, BC = 2230.
$\angle BAC = 2\alpha = 80.94°$.

Problem 1 of the British Mathematical Olympiad 2007

Find the minimum value of $x^2+y^2+z^2$ where x, y, z are real numbers such that $x^3 + y^3 + z^3 - 3xyz = 1$.

Solution

We have

$x^3 + y^3 + z^3 - 3xyz = (x + y + z)(x^2+y^2+z^2 - xy - xz - yz)$ (i)

and $(x + y + z)^2 = x^2+y^2+z^2 + 2xy +2xz + 2yz$, or

$x^3 + y^3 + z^3 - 3xyz = (x^2+y^2+z^2 - xy - xz - yz)\sqrt{x^2+y^2+z^2 + 2(xy +xz + yz)} = 1$, or

$(x^2+y^2+z^2 - xy - xz - yz) = 1/\sqrt{x^2+y^2+z^2 + 2(xy +xz + yz)}$

Now square both sides, we have

$(x^2+y^2+z^2 - xy - xz - yz)^2 = 1/[x^2+y^2+z^2 + 2(xy +xz + yz)]$ (ii)

But from (i), $(x + y + z)(x^2+y^2+z^2 - xy - xz - yz) = 1$ or

$xy + xz + yz = x^2+y^2+z^2 - 1/(x + y + z)$

Substituting $xy + xz + yz$ into (ii), we have

$1/(x + y + z)^2 = 1/\{x^2+y^2+z^2 + 2[x^2+y^2+z^2 - 1/(x + y + z)]\}$, or

$(x + y + z)^2 = 3(x^2+y^2+z^2) - 2/(x + y + z)$

$x^2+y^2+z^2 = 1/3\ [2/(x + y + z) + (x + y + z)^2\ ]$ (iii)

Now let $w = x + y + z$

$x^2+y^2+z^2 = 1/3\ (2/w + w^2)$

$(x^2+y^2+z^2)' = 2/3\ [w - 1/(w^2)]$

$x^2+y^2+z^2$ is minimum when $w - 1/(w^2) = 0$, or when $w^3 = 1$ or $w = 1$

Substituting $w = 1$ into (iii), we have $x^2 + y^2 + z^2 = 1$.

Problem 1 of Romanian Mathematical Olympiad 2006

Let ABC be a triangle and the points M and N on the sides AB and BC, respectively, such that 2CN/BC = AM/AB. Let P be a point on the line AC. Prove that the lines MN and NP are perpendicular if and only if PN is the interior angle bisector of $\angle$MPC.

Solution

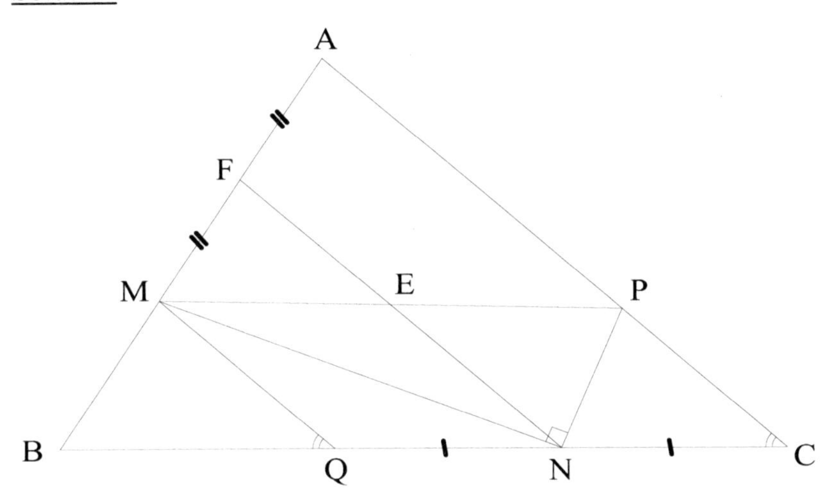

1) Assume MN and NP are perpendicular

Since 2CN/BC = AM/AB, pick point Q on BC such that QN = CN and MQ || AC. Let E and F be the midpoints of MP and MA, respectively.

We have EF || AC but N is also the midpoint of QC and MQ || AC; therefore,, FN || AC and F, E and N are collinear.
We then have $\angle$ENP = $\angle$NPC .

But EN = EP = EM (E is midpoint of MP and $\angle$MNP is right angle) causing $\angle$ENP = $\angle$EPN, or $\angle$EPN = $\angle$NPC, and PN is the interior bisector of $\angle$MPC.

2) Assume PN is interior bisector of $\angle$MPC

$\angle$EPN = $\angle$NPC and since MQ || AC and F and N are the midpoints of MA and QC, respectively, we have FN || AC.

Therefore, $\angle$FNP = $\angle$NPC and E is the midpoint of MP. It follows that $\angle$EPN = $\angle$ENP and EN = EP = EM or $\angle$MNP = 90° and MN is orthogonal to NP.

Problem 1 of USA Mathematical Olympiad 1973

Two points, P and Q, lie in the interior of a regular tetrahedron ABCD. Prove that angle PAQ < 60°.

Solution

Let the side length of the regular tetrahedron be a. Link and extend AP to meet the plane containing triangle BCD at E; link AQ and extend it to meet the same plane at F. We know that E and F are inside triangle BCD and that $\angle PAQ = \angle EAF$

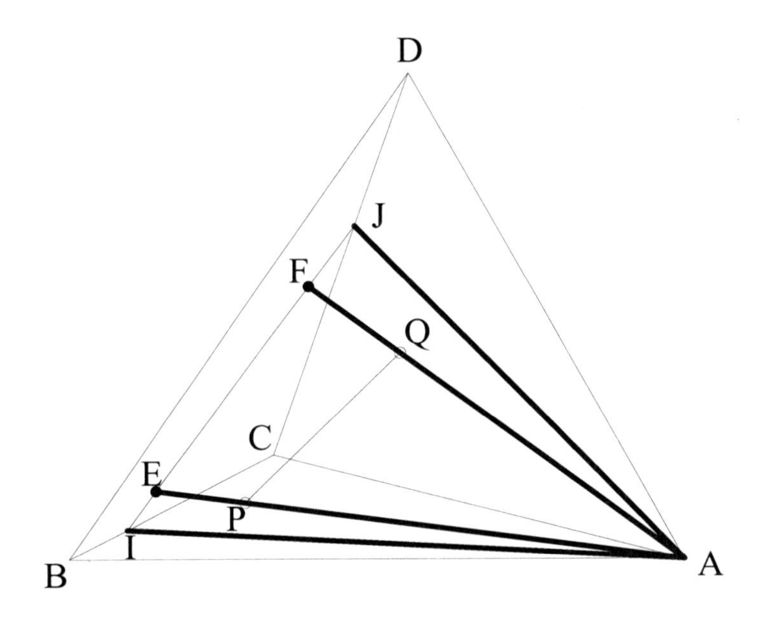

Now let's look at the plane containing triangle BCD with points E and F inside the triangle. Link and extend EF on both sides to meet the sides of the triangle BCD at I and J, I on BC and J on DC. We have $\angle EAF < \angle IAJ$.
But since E and F are interior of the tetrahedron, points I and J cannot be both at the vertices and IJ < a, $\angle IAJ < \angle BAD = 60°$.

Therefore, $\angle PAQ < 60°$.

Problem 1 of USA Mathematical Olympiad 1984

The product of two of the four roots of the quartic equation $x^4 - 18x^3 + kx^2 + 200x - 1984 = 0$ is -32. Determine the value of k.

Solution

The equation having four roots can be expressed as $(x - a)(x - b)(x - c)(x - d) = 0$ where a, b, c and d are the roots. In our case, without loss of generality, we assume the product ab = -32.

Expanding the equation, we have

$$x^4 - (a+b+c+d)x^3 + [ab + cd + (a+b)(c+d)]x^2 + abcd = 0$$

Now equating the corresponding terms, we have

$$a + b + c + d = 18 \qquad \text{(i)}$$
$$ab(c + d) + cd(a + b) = -200 \qquad \text{(ii)}$$
$$abcd = -1984 \qquad \text{(iii)}$$
$$k = ab + cd + (a + b)(c + d) \qquad \text{(iv)}$$

Since ab = -32, from (iii), cd = 62. Let y = a + b and z = c + d, equations (i) and (ii) become

$$y + z = 18$$
$$62y - 32z = -200$$

Solving them we have y = 14 and z = 4. Substituting these values to (iv), k = 86.

Problem 1 of the USA Mathematical Olympiad 2010

Let AXYZB be a convex pentagon inscribed in a semicircle of diameter AB. Denote by P, Q, R, S the feet of the perpendiculars from Y onto lines AX, BX, AZ, BZ, respectively. Prove that the acute angle formed by lines PQ and RS is half the size of ∠XOZ, where O is the midpoint of segment AB.

Solution

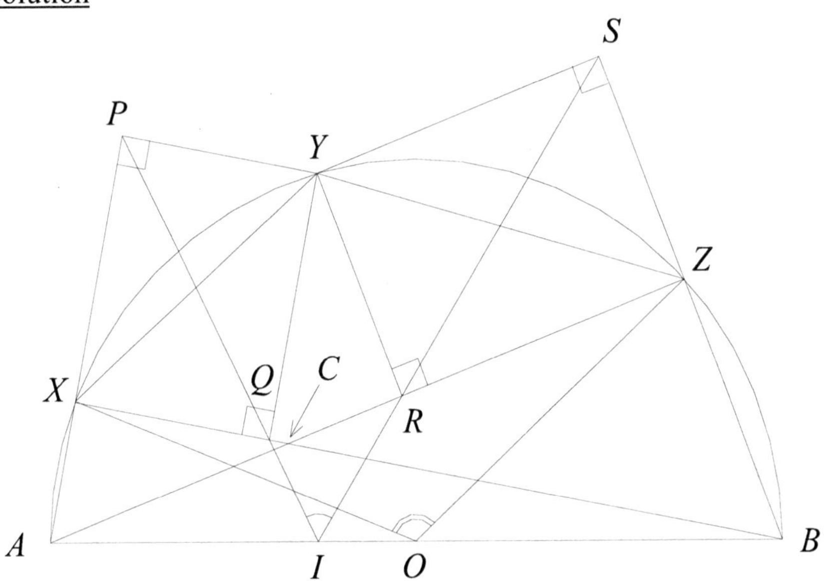

Let AZ intercept BX at C, PQ and RS intercept at I. The acute angle formed by lines PQ and RS is ∠PIS = ∠PQY + ∠SRY − ∠QYR = ∠PQY + ∠SRY − (180° − ∠QCR) = ∠PQY + ∠SRY − ∠RCB.

But ∠RCB subtends arcs AX and BZ; ∠PQY = ∠PXY subtends arc AY; ∠SRY = ∠SZY subtends arc BY.

Therefore, ∠PIS subtends arc AY + BY − AX − BZ = arc XZ = ½ ∠XOZ.

Problem 1 of the William Lowell Putnam Competition 1986

Find, with explanation, the maximum value of $f(x) = x^3 - 3x$ on the set of all real numbers x satisfying $x^4 + 36 \le 13x^2$.

Solution

The maximum value of $f(x) = x^3 - 3x = x(x^2 - 3)$ occurs when x is maximum and it's maximum on the set of all real numbers x satisfying $x^4 + 36 \le 13x^2$ is when x is maximum and still satisfies the inequality $x^4 - 13x^2 + 36 \le 0$.

The localized extreme values of $x^4 - 13x^2 + 36$ occurs when $(x^4 - 13x^2 + 36)' = 0$ or when $4x^3 - 26x = 0$, or when $x = 0$ and $x = \pm\frac{1}{2}\sqrt{26}$.

Now let $y = x^2$, $x^4 - 13x^2 + 36 = y^2 - 13y + 36 = 0$ when

$y = x^2 = 9$ or 4, or $\quad x = \pm 3, \pm 2$

As $x > 3$, $f(x) > 0$ so the maximum value of x to satisfy $x^4 - 13x^2 + 36 \le 0$ is $x = 3$. Substituting it to $x^3 - 3x$, we have $f(3) = 18$.

Problem 2 of the Austrian Mathematical Olympiad 2005

For how many integer values a with $|a| \leq 2005$ does the system of equations

$$x^2 = y + a$$
$$y^2 = x + a$$

have integer solutions?

Solution

Subtracting the two equations, we have $x^2 - y^2 = y - x$
a) When $y \neq x$, we can write $(x + y)(x - y) = y - x$ or $x = -y - 1$, so now we know that if solution x is an integer, y will also be an integer.

Now substituting $x = -y - 1$ into $y^2 = x + a$, we have $y^2 + y + 1 - a = 0$ which has roots as $y = ½ (-1 \pm \sqrt{4a - 3})$ (i)
y has real solutions when $4a - 3 \geq 0$, and it has integer solution when $4a - 3 = m^2$ where m is an integer.

Since $|a| \leq 2005$, $-2005 \leq a \leq 2005$ and $0 \leq 4a - 3 \leq 8017$ (ii)

Values of integers m to satisfy (ii) are $0 \leq m \leq 89$, or the values for $4a - 3$ are $1^2, 3^2, 5^2, 7^2, \ldots, 89^2$. Among these values we have to find the squares that makes a an integer. Let m = pq where q is the units digit. We have

$$a = (100p^2 + q^2 + 20pq + 3)/4$$

Note that both $100p^2$ and $20pq$ are divisible by 4; therefore, $q^2 + 3$ has to be divisible by 4. or when units digit q = 1, 3, 5, 7 or 9. So all the squares of the odd numbers from 1 to 89 will make a an integer and $\sqrt{4a - 3}$ an odd number which, in turn, makes y in (i) an integer. That's a total of 45 numbers for a.

b) When $y = x$, substituting it into the second equation, we have
$y^2 - y - a = 0$

which has roots as $y = ½\,(1 \pm \sqrt{4a + 1}\,)$

y has real solutions when $4a + 1 \geq 0$, and it has integer solution when $4a + 1 = n^2$ where n is an integer.

Since $|a| \leq 2005$, $-2005 \leq a \leq 2005$, and $0 \leq 4a + 1 \leq 8021$ (iii)

Similarly, values of integers n to satisfy (iii) are $0 \leq n \leq 89$, or the values for $4a + 1$ are $1^2, 3^2, 5^2, 7^2, \ldots, 89^2$. Among these values we have to find the squares that makes a an integer. Let n = pq where q is the units digit. We have

$$a = (100p^2 + q^2 + 20pq - 1)/4 = (100p^2 + q^2 + 20pq - 4 + 3)/4$$

Note that $100p^2$, $20pq$ and -4 are divisible by 4; therefore, $q^2 + 3$ has to be divisible by 4 which ends up with the number of integer a being the same as above, 45 of them.

Problem 2 Asian Pacific Mathematical Olympiad 1992

In a circle C with center O and radius r, let $C1$, $C2$ be two circles with centers $O1$, $O2$ and radii $r1$, $r2$ respectively, so that each circle Ci is internally tangent to C at Ai and so that $C1$, $C2$ are externally tangent to each other at A. Prove that the three lines OA, $O1A2$, and $O2A1$ are concurrent.

Solution

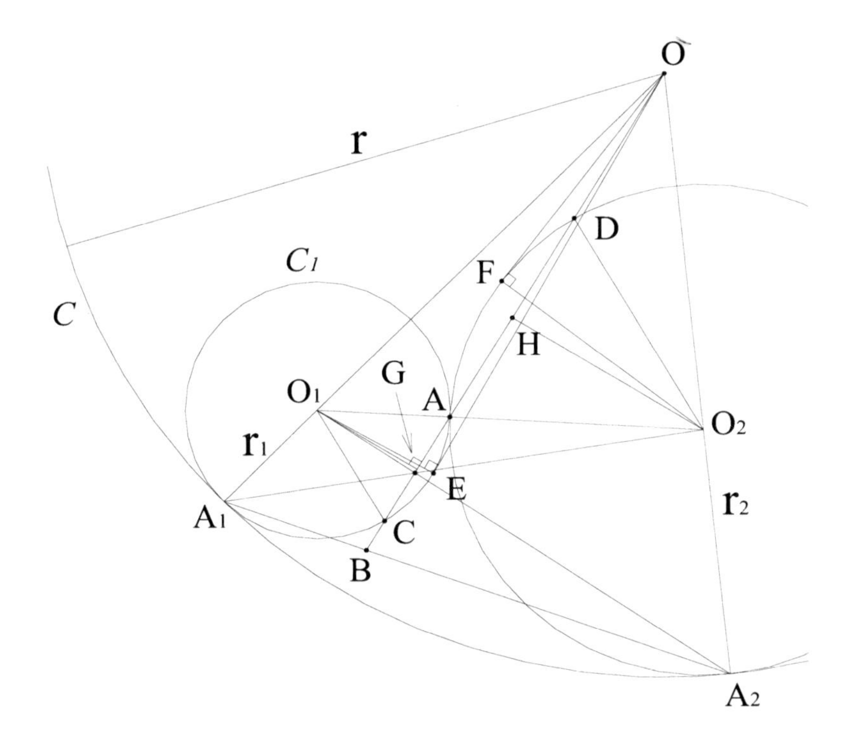

Extend OA to meet A1A2 at B. From O1 and O2 draw altitudes O1G and O2H to OB, respectively. From O draw tangential lines to C1 and C2 and meet them at E and F, respectively.

Use the law of the sine function, we have

$$\frac{A1B}{\sin\angle A1OB} = \frac{OB}{\sin\angle OA1B} = \frac{OB}{\sin\angle OA2B} = \frac{A2B}{\sin\angle A2OB}$$

or

$$\frac{A1B}{A2B} = \frac{\sin\angle A1OB}{\sin\angle A2OB} = \frac{O1G}{O1O} / \frac{O2H}{O2O} = \frac{O1G}{O1O} \times \frac{O2O}{O2H} \qquad \text{(i)}$$

Because the two triangles GO1A and HO2A are similar, we have

$$\frac{O1G}{O1A} = \frac{O2H}{O2A} \qquad \text{or} \qquad \frac{O1G}{O2H} = \frac{r1}{r2} \qquad \text{and (i) becomes}$$

$$\frac{A1B}{A2B} = \frac{r1}{r2} \times \frac{O2O}{O1O} = \frac{r1}{r2} \times \frac{r - r2}{r - r1}, \qquad \text{or} \qquad \frac{A1B}{A2B} \times \frac{r2}{r - r2} \times \frac{r - r1}{r1} = 1$$

Therefore, per Ceva's theorem, the three lines OA, O1A2, and O2A1 are concurrent.

Problem 2 Asian Pacific Mathematical Olympiad 2003

Suppose ABCD is a square piece of cardboard with side length a. On a plane are two parallel lines l1 and l2, which are also a units apart. The square ABCD is placed on the plane so that sides AB and AD intersect l1 at E and F respectively. Also, sides CB and CD intersect l2 at G and H respectively. Let the perimeters of triangle AEF and triangle CGH be m1 and m2 respectively. Prove that no matter how the square was placed, m1 + m2 remains constant.

Solution

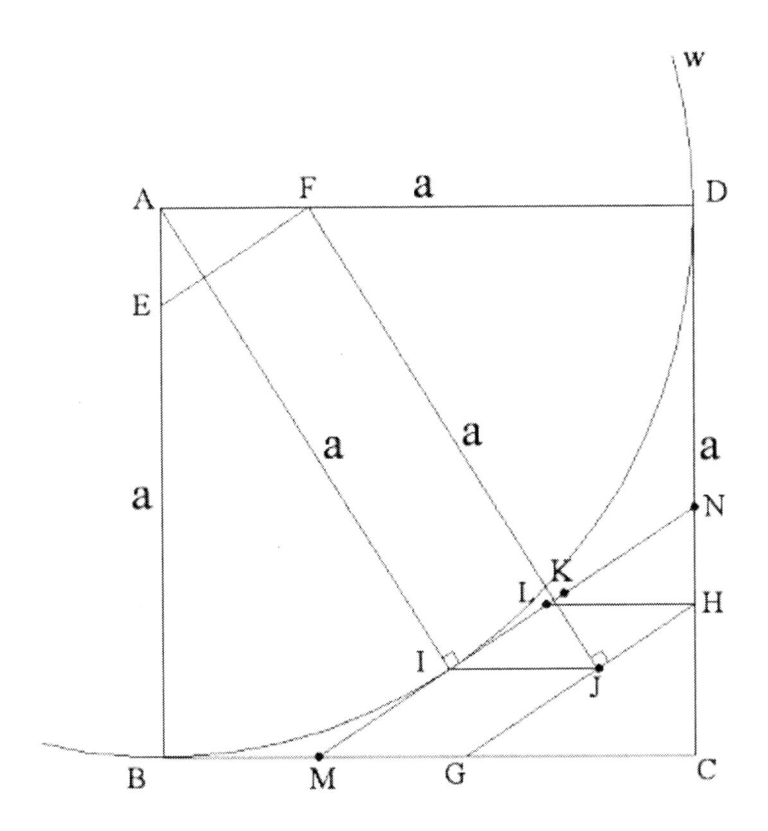

It's easily seen that the two triangles AEF and CHG are similar. We have

HC / AE = GC / AF.

Picks points M and N on BC and DC, respectively such that NH = AE and MG = AF.

We then have HC / NH = GC / MG, or GH || MN, and

From H draw a line parallel to AD and intercept MN at L. Triangles AEF and HNL are congruent. Therefore,

AE = NH, AF = LH = MG, EF = LN, GH = ML

m1 = AE + AF + EF
m2 = HC + GC + GH

m1 + m2 = AE + AF + EF + HC + GC + GH = NH + HC + GC + MG + ML + LN = NC + MC + MN

or m1 + m2 is the perimeter of triangle MCN.

From F draw a line perpendicular to and intercept GH at J, we have FJ = a as given by the problem. Similarly, from A draw a line perpendicular to and intercept MN at I, we have

FJ = AI = a

That proves to us that line MN is tangential to the circle with radius a and center A. Therefore, the parameter of triangle MCN equals BC + DC = 2a, or m1 + m2 is a constant.

<u>Further observation</u>

Let K be the foot of incenter of incircle of triangle MCN to MN. Prove that IK = MN – 2 × KN.

Problem 2 of Asian Pacific Mathematical Olympia 2004

Let O be the circumcenter and H the orthocenter of an acute triangle ABC. Prove that the area of one of the triangles AOH, BOH and COH is equal to the sum of the areas of the other two.

Solution 1

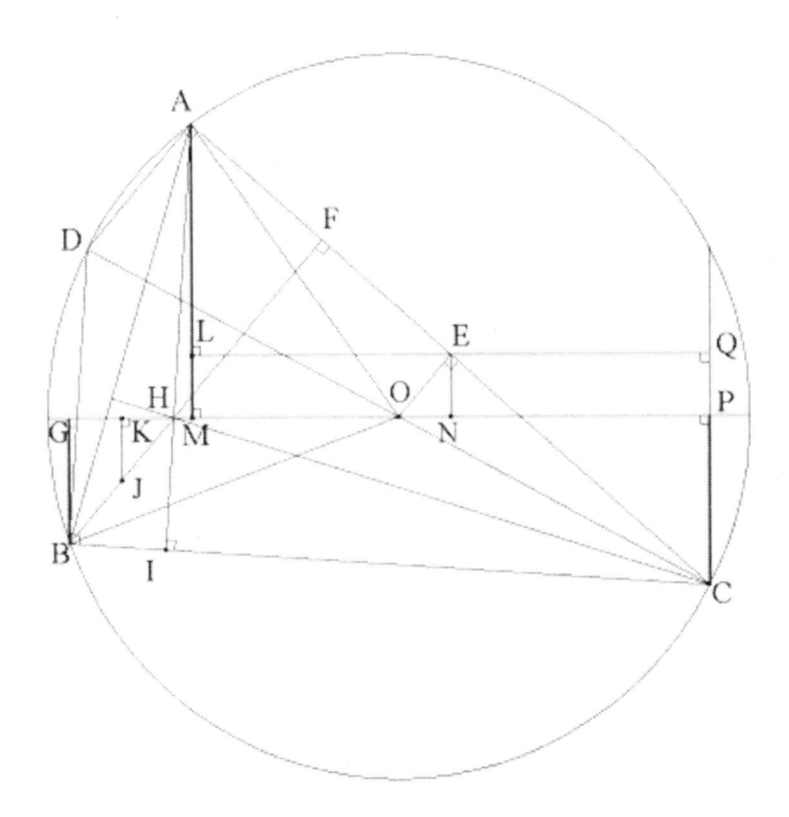

From the vertices of triangle ABC A, B and C draw the altitudes to OH and intercept the extension of OH at M, G and P, respectively.

Since the three triangles AOH, BOH and COH have the same base OH, to prove the area of AOH is the sum of the areas BOH and COH, we need to prove

$$AM = GB + PC \qquad \text{(i)}$$

Extend CO to intercept the circle at D. Since CD is the diameter of the circle, we have $\angle DAC = \angle DBC = \angle BFC = \angle AIC = 90$ degrees.

Or $AD \parallel HB$ and $DB \parallel AH$; therefore, $AD = HB$

O and E are also midpoints of DC and AC, respectively,
we have $OE = \frac{1}{2} AD$

Or $OE = \frac{1}{2} HB$

Let J be the midpoint of BH; from J draw the altitude to OH and cuts the extension of OH at K. We have

$$KJ = \tfrac{1}{2} GB \qquad \text{(ii)}$$
$$HJ = \tfrac{1}{2} HB = OE \text{ and}$$
$$\angle KHJ = \angle OHF = \angle NOE$$

From E draw the altitude to OH and intercept it at N. The two triangles JKH and ENO are then congruent; we then have

$KJ = EN$

Draw the line parallel to OH through E and intercepts AM and PC at L and Q, respectively; we then have

$$KJ = EN = LM = QP$$
$$AL = QC \qquad \text{(iii)}$$

Combining with (ii) $\quad GB + PC = 2\,KJ + QC - QP$

From (iii) $\quad GB + PC = LM + QP + AL - QP$

Or $\quad GB + PC = AM$

which is the condition (i) we set out to prove.

Solution 2

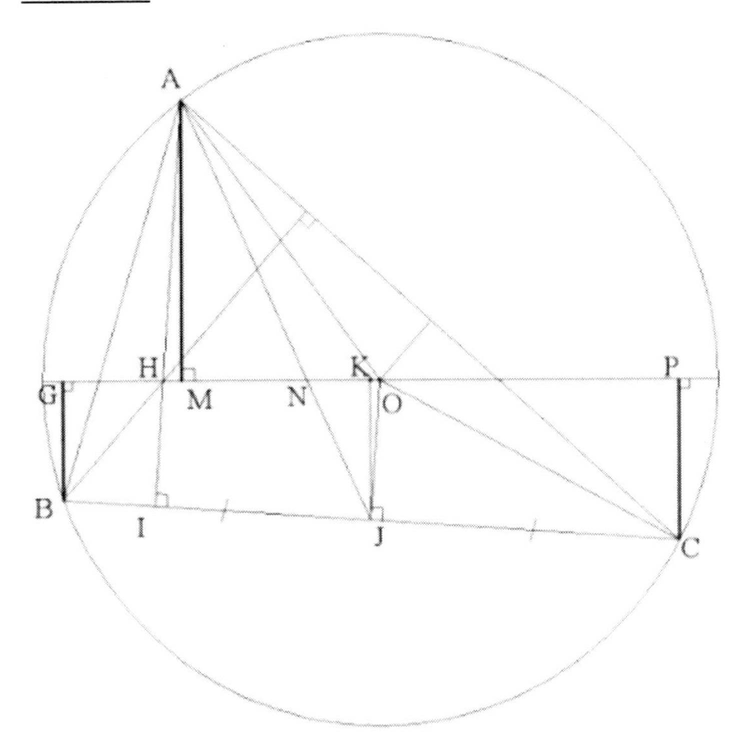

From the three vertices A, B and C of ΔABC draw orthogonal lines to OH and intercept it at M, G and P, respectively. The three triangles AOH, BOH and COH share the same base OH, so to prove the areas of AOH to equal the areas of the other two we need to prove AM = GB + PC.

Let J be the midpoint of BC. AJ intercepts OH at N. From J draw the line to ⊥ and intercept OH at K. We see that GB + PC = 2JK. We then need to prove AM = 2JK. Note that in a triangle, the three points centroid, orthocenter and circumcenter collinear; therefore, N is also the centroid of ΔABC

Therefore, AN = 2NJ.

or AM = 2JK because the two triangles AMN and JKN are similar.

Problem 2 of the Canadian Mathematical Olympiad 1977

Let O be the center of a circle and A a fixed interior point of the circle different from O. Determine all points P on the circumference of the circle such that the angle OPA is a maximum.

Solution

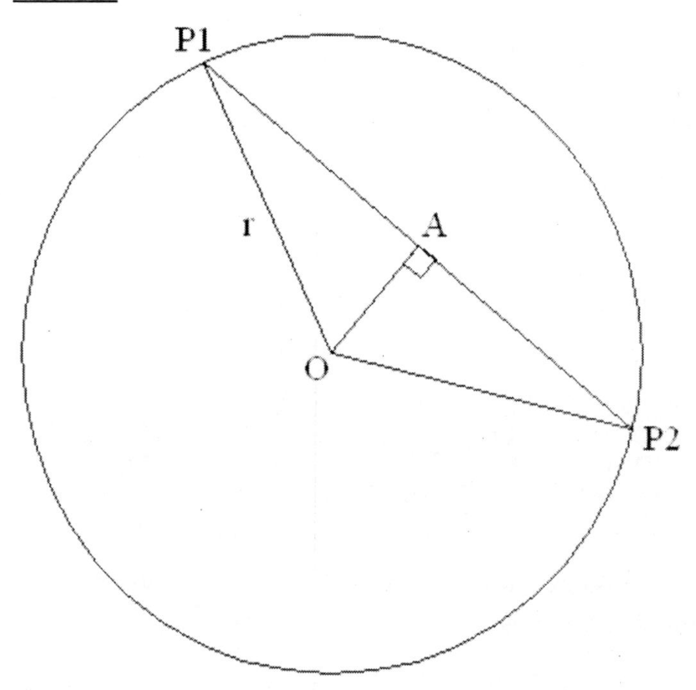

Let point P = point P1. Apply the law of the sine function, we obtain $OA / \sin\angle OP1A = r / \sin\angle OAP1$, or $\sin\angle OP1A = OA \sin\angle OAP1 / r$. Since angle OP1A has a side passing through the center of the circle, it's an acute angle, and therefore, the angle $\angle OP1A$ is maximum when $\sin\angle OP1A$ is maximum. Furthermore, since r and OA are constants, $\sin\angle OP1A$ is maximum when $\sin\angle OAP1$ is maximum, and the maximum of a sine of an angle is 1 which will happen when $\angle OAP1 = 90°$.

Another point P = P2 which is the mirror image of P1 across A also satisfies this requirement.

Problem 2 of the Canadian Mathematical Olympiad 1978

Find all pairs a, b of positive integers satisfying the equation
$2a^2 = 3b^3$.

Solution

The product on the left side $2a^2$ is an even number, so $3b^3$ has to be an even number, and b^3; therefore,, has to be an even number, or b to be an even number. Let $b = 2n$ where n is a positive integer.

We then have $b^2 = 4n^2$; now rewrite $2a^2 = 3b^3$ as $2a^2 / b^2 = 3$, or

$2a^2 / (4n^2) = 3b$, or $a^2 / (2n^2) = 3b$, or $a^2 = 6bn^2$

Since a^2 and n^2 are already squares of two numbers, we have to have $6b$ to be the square of another number. Let it be $6b = m^2$, or

$b = 6k^2$ where k is a positive integer. Now substituting it to the original equation, we have

$a^2 = 3b^3 /2 = 3^2x6^2(k^3)^2$ or $a = 18k^3$

Solutions are $(a, b) = (18k^3, 6k^2)$ where k is a positive integer. For example, for $k = 23$, $a = 18 \times 23^3 = 219006$ and $b = 6 \times 23^2 = 3174$ is a set of solution when

$2 \times 219006^2 = 3 \times 3174^3 = 95{,}927{,}256{,}072$

Problem 2 of the Canadian Mathematical Olympiad 1981

Given a circle of radius r and a tangent line *l* to the circle through a given point P on the circle. From a variable point R on the circle, a perpendicular RQ is drawn to *l* with Q on *l*. Determine the maximum of the area of triangle PQR.

Solution (This solution applies college calculus)

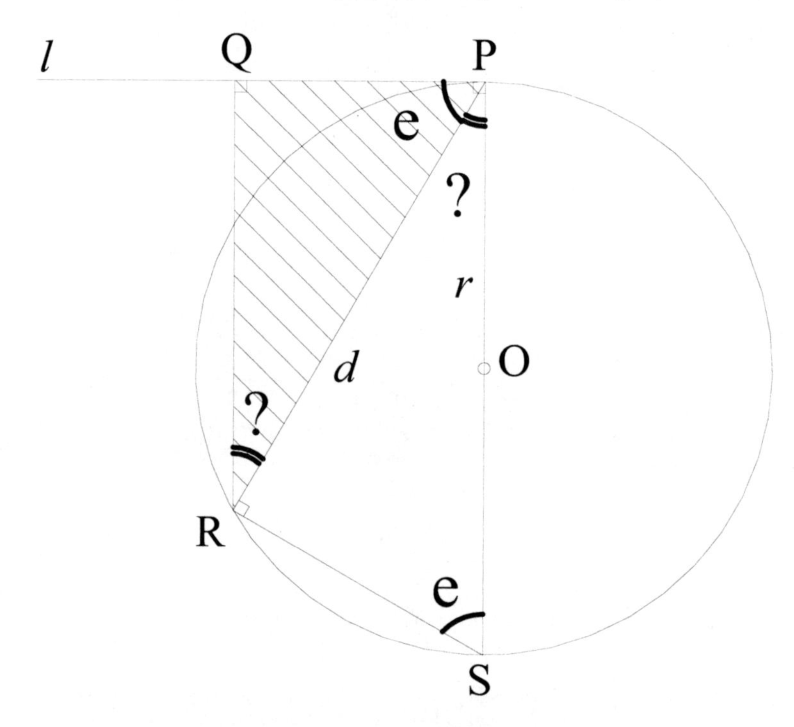

Let O be the center of the circle. Link and extend PO to meet the circle at S. Now let $\varepsilon = \angle QPR$ and $\theta = \angle QRP$. We then also have

$\varepsilon = \angle PSR$ and $\theta = \angle RPS$. Let d = PR and denote (Ω) the area of shape Ω. We have (PQR) = ½ PQ × RQ

But PQ = d sinθ, and RQ = d sinε, and

$$(PQR) = \tfrac{1}{2} d^2 \sin\theta \sin\varepsilon \qquad \text{(i)}$$

But in triangle PRS, $\sin\varepsilon = d / (2r)$ and $\sin\theta = RS / (2r)$

Substituting them into (i), we have

$$(PQR) = d^3 RS / (8r^2) \qquad \text{(ii)}$$

With radius r being a constant, to find the maximum value of (PQR) we now need to find the maximum value of function $f(d) = d^3 RS$, and we're stuck. But there's a trick. Recall that a function reaches its extreme (maximum or minimum) point when its derivative is zero?

But the function f(d) above has two variables d and RS. Now let's try to reduce it to a single variable d by relating RS to variable d

and to eliminate RS. We have $RS = \sqrt{4r^2 - d^2}$ so now $f(d) = d^3 \sqrt{4r^2 - d^2}$ and f'(d) is the derivative of f(d) with respect to the changing variable d, and it is the derivative of the product of two differentiable functions. We have the formula

$$Dx(u . v) = u . Dxv + v . Dxu$$

Therefore, $f'(d) = [d^3 \sqrt{4r^2 - d^2}\,]' = d^3 [\sqrt{4r^2 - d^2}\,]' + [\sqrt{4r^2 - d^2}\,] (d^3)'$, but we also have

Dx(x to power of n) = n (x to power of n – 1) and now

$$f'(d) = [\,½d^3/\sqrt{4r^2 - d^2}\,]\, Dd(4r^2 - d^2) + 3d^2\sqrt{4r^2 - d^2} =$$

$$[½d^3/\sqrt{4r^2 - d^2}\,](0-2d) + 3d^2\sqrt{4r^2 - d^2} =$$

$$-d^4/\sqrt{4r^2 - d^2} + 3d^2\sqrt{4r^2 - d^2}$$

The derivative $f'(d) = 0$ when $d^4/\sqrt{4r^2 - d^2} = 3d^2\sqrt{4r^2 - d^2}$ or when $d^2 = 3(4r^2 - d^2)$

or when $d^2 = 3r^2$ or $d = r\sqrt{3}$.

We know the minimum of (PQR) occurs when it's a degenerate triangle either by having R at P ($R \equiv P$ and $d = 0$) or R at S ($R \equiv S$

and $d = 2r$) and $(PQR) = 0$. Neither is the case when $d = r\sqrt{3}$ when (PQR) is maximum.

When $d = r\sqrt{3}$, $\sin\varepsilon = d / (2r) = \sqrt{3}/2$ or $\varepsilon = 60°$ as seen on the graph, and

max $(PQR) = d^3 \sqrt{4r^2 - d^2} / (8r^2) = (r\sqrt{3})^3 \sqrt{4r^2 - d^2} / \sqrt{8r^2} =$

$3\sqrt{3}\, r^2/8$.

Problem 2 of Canadian Mathematical Olympiad 1985

Prove or disprove that there exists an integer which is doubled when the initial digit is transferred to the end.

Solution

Assume that there is such an integer N. Let $N = n_0 n_1 n_2 \ldots$

$n_{n-1} n_n$ $(n_0 \neq 0)$ and $2\, n_0 n_1 n_2 \ldots n_{n-1} n_n = n_1 n_2 \ldots$

$n_{n-1} n_n n_0$

Since the number on the left is even, the units digit of the number on the right must also be even, or n_0 is an even digit.

Expanding the above equation, we have

$2n_0 \times 10^n + 2n_1 \times 10^{n-1} + 2n_2 \times 10^{n-2} + \ldots + 2n_{n-1} \times 10 + 2n_n$

$= n_1 \times 10^n + n_2 \times 10^{n-1} + \ldots + n_{n-1} \times 102 + n_n \times 10 + n_0$

Now regroup them all

$n_0 (2 \times 10^n - 1) - 8n_1 \times 10^{n-1} - 8n_2 \times 10^{n-2} - \ldots - 8n_{n-1} \times 10$

$-8n_n = 0$ (i)

Since n_0 is even, $n_0 = 2, 4, 6$ or 8

When $n_0 = 2$, divide the left side of (i) by 2, we have

$2 \times 10^n - 1 - 4n_1 \times 10^{n-1} - 4n_2 \times 10^{n-2} - \ldots - 4n_{n-1} \times 10 - 4n_n =$

0.

We see that the left side is now an odd number which is not zero, so $n_0 \neq 2$.

When $n_0 = 4$, dividing the left side of (i) by 4, we have

$2 \times 10^n - 1 - 2n_1 \times 10^{n-1} - 2n_2 \times 10^{n-2} - \ldots - 2n_{n-1} \times 10 - 2n_n = 0.$

Again the left side is an odd number and not zero, so $n_0 \neq 4$

When $n_0 = 6$ or 8 the second number

$n_1 n_2 \ldots\ldots n_{n-1} n_n n_0$ has one more digit than the first number

$n_0 n_1 n_2 \ldots n_{n-1} n_n$ so $n_0 \neq 6$ and $n_0 \neq 8$.

Conclusion

There exists no integer which is doubled when the initial digit is transferred to the end.

Problem 2 of Canadian Mathematical Olympiad 1987

The number 1987 can be written as a three digit number xyz in some base b. If $x + y + z = 1 + 9 + 8 + 7$, determine all possible values of x, y, z, b.

Solution

Converting the number xyz in base b to base 10, we have

$$xb^2 + yb + z = 1987 \quad \text{but}$$
$$x + y + z = 1 + 9 + 8 + 7 = 25$$

subtracting the two equations, we have

$(b - 1)[x(b + 1) + y] = 1987 - 25 = 1962 = 2 \times 3 \times 3 \times 109$
where all these numbers are prime.

We can only find solution of $b - 1 = 19 - 1$ and $x(b + 1) + y = 5 \times (19 + 1) + 9$.

and $z = 25 - 5 - 9 = 11$ (or $z = B$)

Answer: $x = 5$, $y = 9$, $z = B$, and $b = 19$.

Problem 2 of the Canadian Mathematical Olympiad 1988

A house is in the shape of a triangle, perimeter P meters and area A square meters. The garden consists of all the land within 5 meters of the house. How much land do the garden and house together occupy?

Solution

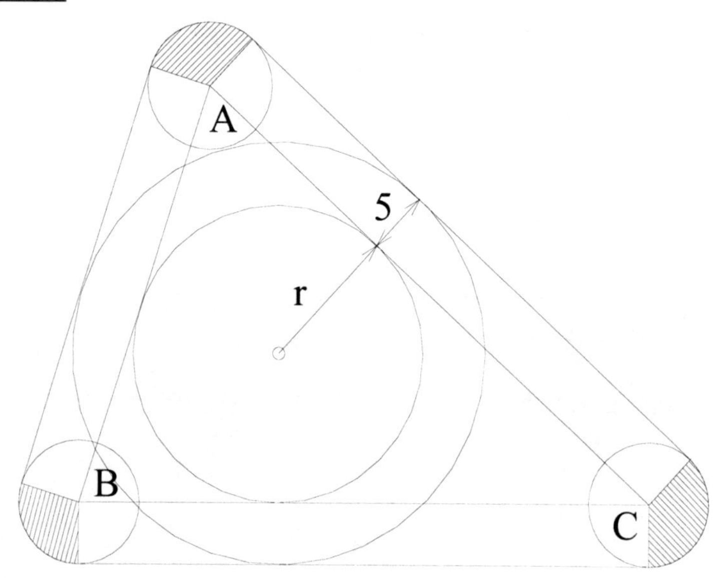

The garden and the house occupy the area bounded by the outside perimeter of the picture as shown.

Easily note that the area of the three shaded areas combined to be the area of a circle with radius of 5 = pi × 5^2 = 78.54.

The areas of the three rectangles is 5xP = 5P. The area of the garden and house together is A + 5P + 78.54 square meters.

Problem 2 of Canadian Mathematical Olympiad 1989

Let ABC be a right angled triangle of area 1. Let A'B'C' be the points obtained by reflecting A, B, C respectively, in their opposite sides. Find the area of triangle A'B'C'.

Solution

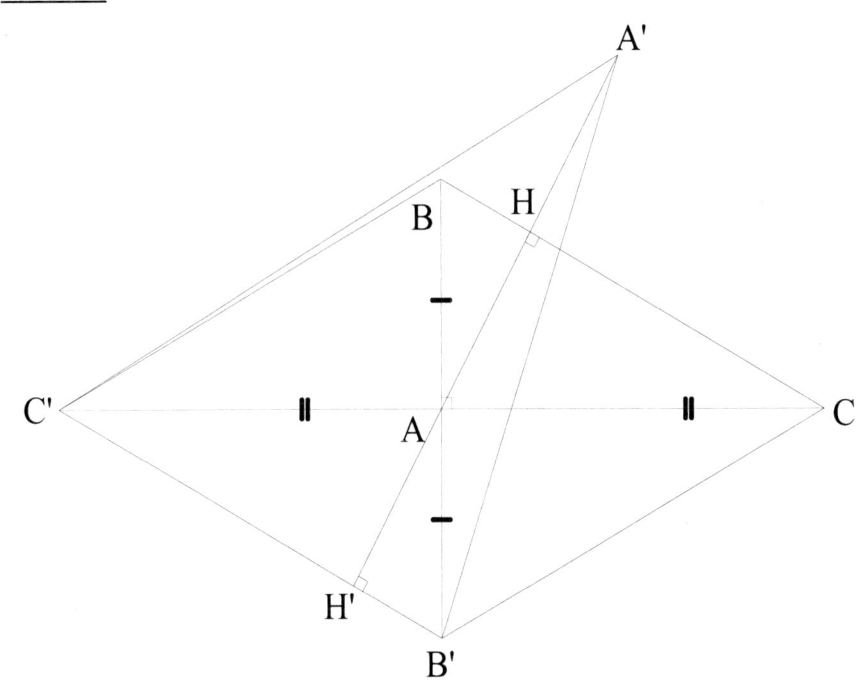

Let (Ω) denote the area of shape Ω. Since A is the midpoint of BB' and CC' and BB' $\perp$ CC', BCB'C' is a rhombus.

Therefore, BC || B'C', BC = B'C' and A'A $\perp$ B'C' and AH = AH' where H' is the foot of A to B'C', and AH = AH' = A'H = 1/3 A'H'.

Therefore, (A'B'C') = ½ A'H' × B'C' = ½ × 3 × AH × BC = 3(ABC) = 3.

Problem 2 of Canadian Mathematical Olympiad 1992

For x, y, z $\geq$ 0, establish the inequality

$x(x - z)^2 + y(y - z)^2 \geq (x - z)(y - z)(x + y - z)$

and determine when equality holds.

Solution

Expanding both sides of the inequality, we obtain
$x^3 - 2x^2z + xz^2 + y^3 - 2y^2z + yz^2 \geq x^2y - x^2z - 3xyz + xz^2 + xy^2 - y^2z + yz^2 + xz^2 + yz^2 - z^3$ or $x^3 + y^3 + z^3 \geq x^2(y + z) + y^2(z + x) + z^2(x + y) - 3xyz$ (i)

We know that $x^3 + y^3 + z^3 - 3xyz = (x + y + z)(x^2 + y^2 + z^2 - xy - xz - yz)$, or
$x^3 + y^3 + z^3 = (x + y + z)(x^2 + y^2 + z^2 - xy - xz - yz) + 3xyz$
The following process is easy to understand:

Now rewrite (i) as
$(x + y + z)(x^2 + y^2 + z^2 - xy - xz - yz) + 3xyz \geq x^2(y + z) + y^2(z + x) + z^2(x + y) - 3xyz$, or
$x^3 + y^3 + z^3 + x^2(y + z) + y^2(z + x) + z^2(x + y) - (x + y + z)(xy + xz + yz) \geq x^2(y + z) + y^2(z + x) + z^2(x + y) - 6xyz$, or
$x^3 + y^3 + z^3 - 3xyz \geq (x + y + z)(xy + xz + yz) - 9xyz$, or
$(x + y + z)(x^2 + y^2 + z^2 - xy - xz - yz) \geq (x + y + z)(xy + xz + yz) - 9xyz$, or

$x^2 + y^2 + z^2 - xy - xz - yz \geq xy + xz + yz - 9xyz/(x + y + z)$, or

$x^2 + y^2 + z^2 - 2xy - 2xz - 2yz \geq -9xyz/(x + y + z)$, or
$(x - y - z)^2 \geq -9xyz/(x + y + z)$

This latest inequality is true since with all x, y, z $\geq$ 0, the left side is non-negative whereas the right side is non-positive. Equality holds when $x = y = z = 0$

Problem 2 of the International Mathematical Olympiad 2007

Consider five points A,B,C,D and E such that ABCD is a parallelogram and BCED is a cyclic quadrilateral. Let l be a line passing through A. Suppose that l intersects the interior of the segment DC at F and intersects line BC at G. Suppose also that EF = EG = EC. Prove that l is the bisector of angle DAB.

Solution

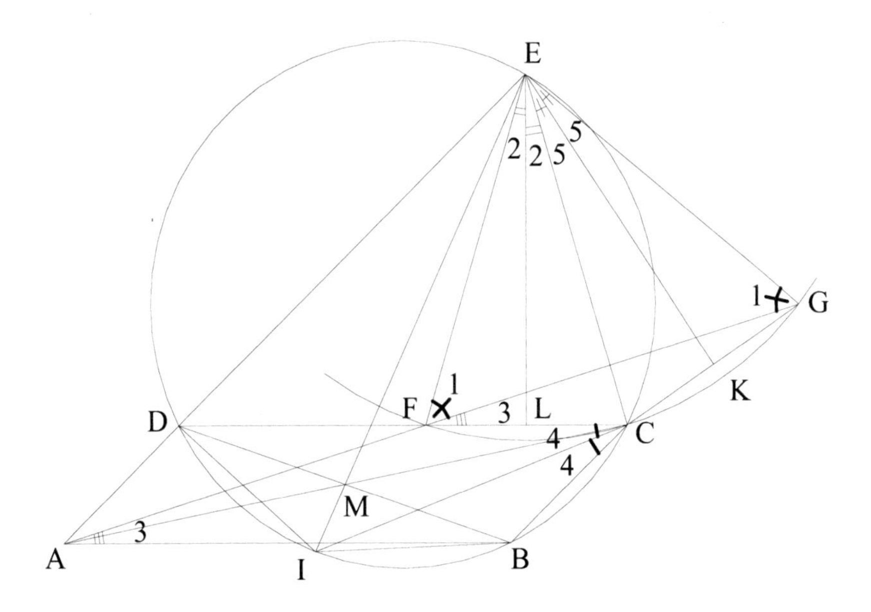

Based on Simson-Wallace's theorem the feet of projections of E down to the three sides of triangle BCD are M, L and K are collinear as seen on the graph.

Since L and K are midpoints of FC and CG, respectively, therefore, LK || FG, and triangle ACG has LK intersect AC at its midpoint M.

Therefore, M is also midpoint of DB since ABCD is a parallelogram.

Thus there is only one unique point M to satisfy conditions that M is on DB and also collinear with K and L, and M is also the foot of E to DB.

Extend EM to cut the circle at I. Since EI is the perpendicular bisector of DB because DM = BM it is understood that I is midpoint of arc DB. Therefore, $\angle DCI = \angle BCI$ = angle $\angle 4$ as denoted on the graph. EI is also the diameter of the circle.

Therefore, $\angle ECI = 90°$

$\angle ICB = \angle CEK$ (they both have sides perpendicular to one another)

or $\angle 4 = \angle 5$ (i)

In triangle EFL : $\angle 1 + \angle 2 + \angle 3 = 90°$

In triangle EFG : $2 \times (\angle 1 + \angle 2 + \angle 5) = 180°$

or $\angle 1 + \angle 2 + \angle 5 = 180°$

Therefore, $\angle 3 = \angle 5$

From (i) $\angle 3 = \angle 4$

Or $\angle 3 = ½ \angle DCB = ½ \angle DAB$ or AG is bisector of $\angle DAB$ which is the answer.

Problem 2 of the International Mathematical Olympiad 2009

Let ABC be a triangle with circumcenter O. The points P and Q are interior points of the sides CA and AB, respectively. Let K, L and M be the midpoints of the segments BP, CQ and PQ, respectively, and let Γ be the circle passing through K, L and M. Suppose that the line PQ is tangent to the circle Γ. Prove that OP = OQ.

Solution

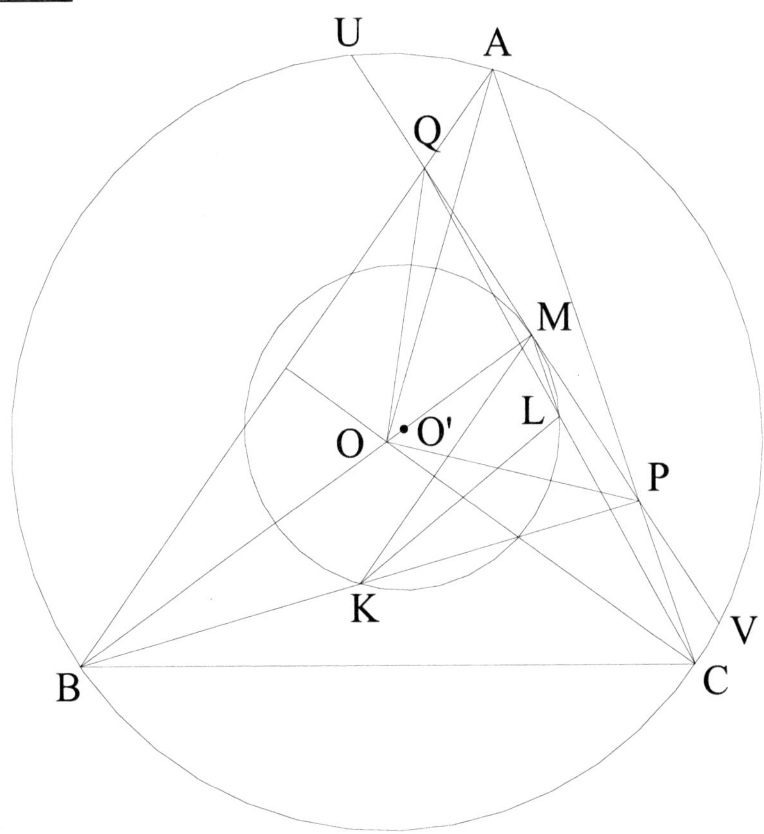

QP tangents with the small circle at M, we have ∠QMK = ∠MLK.

M, K and L are midpoints of PQ, BP and QC, respectively; therefore,

$KM \parallel QB$, $KM = \frac{1}{2} QB$ (i)
$ML \parallel PC$, $ML = \frac{1}{2} PC$ (ii)

and $\angle QMK = \angle MQA$, or $\angle MLK = \angle MQA$.

$ML \parallel PC$ and $KM \parallel QB$ therefore, $\angle QAP = \angle KML$

The two triangles QAP and KML are similar since their respective angles are equal.

Therefore, $ML/QA = KM/AP$

From (i) and (ii) $AP \times PC = QA \times QB$ (iii)

Extend PQ and QP to meet the larger circle at U and V, respectively.

In the larger circle UV intercepts AB at Q, we have

$QU \times QV = QA \times QB$
or $QU \times (QP + PV) = QA \times QB$ (iv)

UV intercepts AC at P, we have
$UP \times PV = AP \times PC$ or
$(QU + QP) \times PV = AP \times PC$ (v)

From (iv) and (iii) $QU \times (QP + PV) = AP \times PC$
Therefore, from (v) $QU \times (QP + PV) = (QU + QP) \times PV$

Or $PV = QU$ and M is also the midpoint of UV and $OM \perp UV$

Therefore, $OP = OQ$.

Problem 2 of the Irish Mathematical Olympiad 2006

P and Q are points on the equal sides AB and AC respectively of an isosceles triangle ABC such that AP = CQ. Moreover, neither P nor Q is a vertex of ABC. Prove that the circumcircle of the triangle APQ passes through the circumcenter of the triangle ABC.

Solution

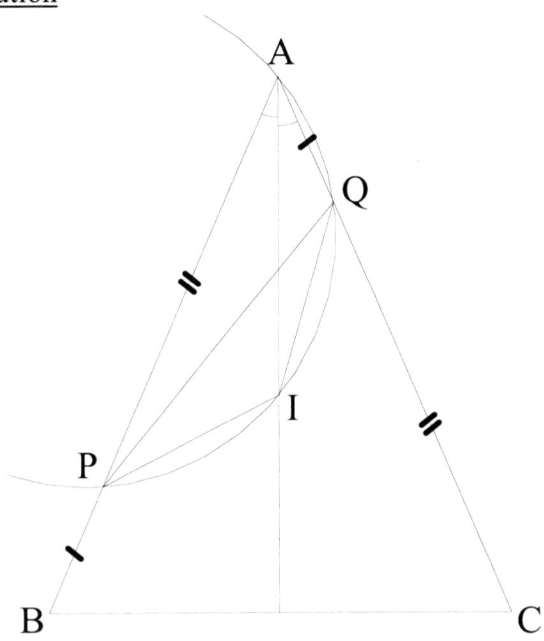

Let the circumcircle of triangle APQ intercept the bisector of $\angle A$ of triangle ABC at I. We have $\angle PAI = \angle QAI$ and PI = QI.

Since AQIP is cyclic, we have $\angle AQI + \angle API = 180°$ or $180° - \angle API = \angle AQI$. Now consider the two triangles BPI and AQI, we have BP = AQ, PI = QI, and $\angle BPI = 180° - \angle API = \angle AQI$ they are congruent and thus AI = BI.

Since AI is the bisector of $\angle BAC$ and ABC is an isosceles triangle with AB = AC, AI is also the altitude to BC, and BI =CI. Therefore, BI = CI = AI or I is the circumcenter of triangle ABC.

Problem 2 of Irish Mathematical Olympiad 2007

Prove that a triangle ABC is right-angled if and only if

$$\sin^2 A + \sin^2 B + \sin^2 C = 2$$

Solution

Let the three side lengths of triangle ABC be a, b and c. Apply the law of the sine function, we obtain

$$\frac{a^2}{\sin^2 A} = \frac{b^2}{\sin^2 B} = \frac{c^2}{\sin^2 C} = \frac{a^2 + b^2 + c^2}{\sin^2 A + \sin^2 B + \sin^2 C} = \frac{a^2 + b^2 + c^2}{2}$$

and the law of the cosines gives us $a^2 = b^2 + c^2 - 2bc\cos A$
Now substituting a^2 into the above equation

$$\frac{a^2}{\sin^2 A} = \frac{a^2 + b^2 + c^2}{2} = \frac{2(b^2 + c^2 - bc\cos A)}{2} = b^2 + c^2 - bc\cos A, \text{ or}$$

$a^2 = (b^2 + c^2 - bc\cos A)\sin^2 A$, or
$b^2 + c^2 - 2bc\cos A = (b^2 + c^2 - bc\cos A)\sin^2 A$, or

$(b^2 + c^2)(1 - \sin^2 A) = bc\cos A(2 - \sin^2 A)$, or

$(b^2 + c^2)\cos^2 A = bc\cos A(1 + \cos^2 A)$, or
$(b^2 + c^2)\cos A = bc(1 + \cos^2 A)$, or
$bc\cos^2 A - (b^2 + c^2)\cos A + bc = 0$

Solving for cosA, we have
$\cos A = b/c$ and c/b; this implies that either angle B or angle C is right.

Now if the triangle is right-angled, we have one of the angle being 90°. Assume it's angle A, and $\sin^2 A = 1$, $\sin^2 B = b^2/a^2$ and $\sin^2 C = c^2/a^2$, and $\sin^2 B + \sin^2 C = (b^2 + c^2)/a^2 = 1$.

Therefore, $\sin^2 A + \sin^2 B + \sin^2 C$.

Problem 2 of the British Mathematical Olympiad 2005

In triangle ABC, $\angle BAC = 120°$. Let the angle bisectors of angles A, B and C meet the opposite sides in D, E and F, respectively. Prove that the circle on diameter EF passes through D.

Solution

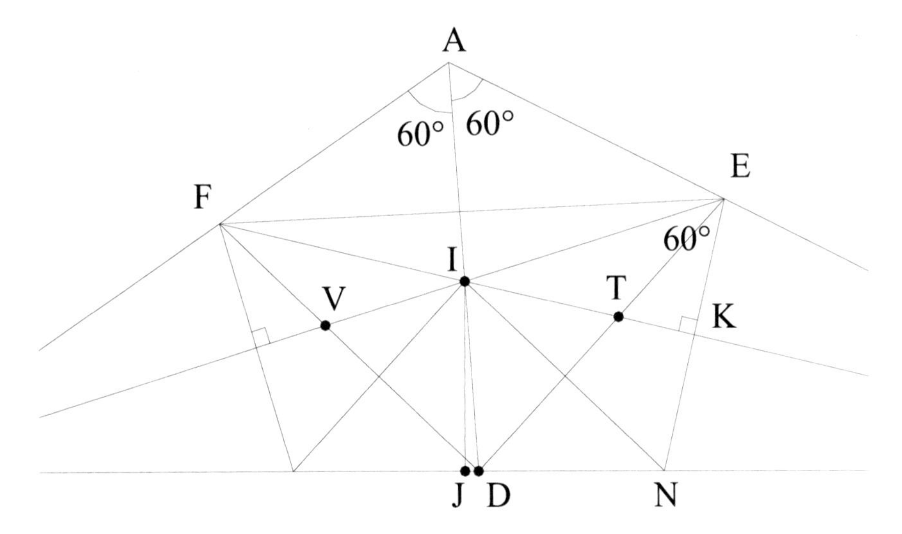

Let BE meet CF at I and J be the foot of I on BC. From E draw the perpendicular to CF to meet CF and BC at K and N, respectively. Also let CF meet ED at T, BE meet FD at V.

We have $\angle BID = \angle ABI + \angle BAI = 90° - ½\angle C = \angle JIC$ or $\angle BIJ = \angle DIC$.

It's easily seen that $\angle EIK = ½ (\angle B + \angle C) = 30°$ or $\angle BIC = 150°$ and $\angle IEK = 90° - \angle EIK = 60°$ and since CI is the perpendicular bisector of EN, IE = IN and $\angle INE = 60°$. It follows that IEN is an equilateral triangle and $\angle NIK = 30°$.

We now have

$\angle IND = \angle NIK + \frac{1}{2}\angle C = 30° + \frac{1}{2}\angle C = 30° + (30° - \frac{1}{2}\angle B) = 60° - \frac{1}{2}\angle B$, and

$\angle DIN = \angle DIC - 30° = \angle BIJ - 30° = 90° - \frac{1}{2}\angle B - 30° = 60° - \frac{1}{2}\angle B$.

or $\angle IND = \angle DIN$ and DE is bisector of $\angle IDN$.

Similarly on the other side, DF is bisector of $\angle IDB$.

Therefore, the angle $\angle FDE = 90°$ and the circle on diameter EF passes through D.

Problem 2 of the British Mathematical Olympiad 2007

Let triangle ABC have incenter I and circumcenter O. Suppose that $\angle AIO = 90°$ and $\angle CIO = 45°$. Find the ratio AB : BC : CA.

Solution

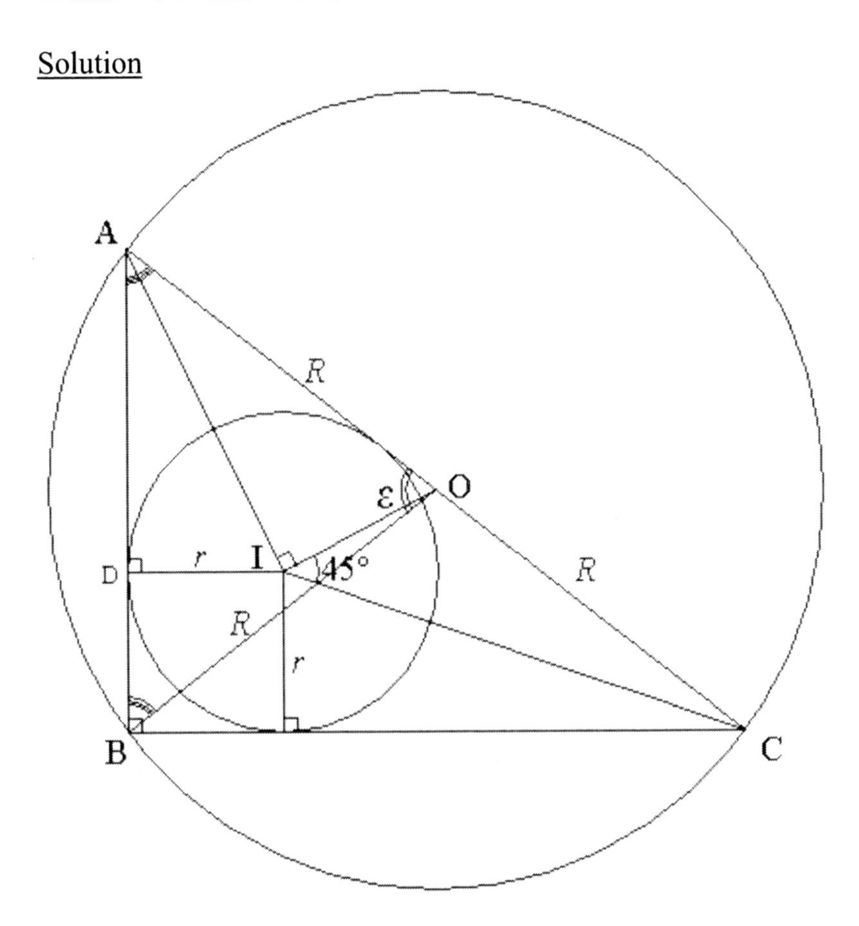

Let the incircle tangent AB at D and r be its radius. We have
$\frac{1}{2}\angle A + \frac{1}{2}\angle C + \angle B = \angle AIC = 90° + 45° = 135°$

and $$\angle A + \angle B + \angle C = 180°$$

or $\angle B = 90°$ and the circumcenter O of triangle ABC is the midpoint of AC.

Now let $\frac{1}{2}AC = OA = OC = OB = R$, the radius of the circumcircle, $OI = b$ and $\angle AOB = \varepsilon$

Apply the law of the sine function for triangle OIC, we obtain

$R/\sin 45° = b/\sin\frac{1}{2}\angle C$, but in triangle AOI, $R = b/\sin\frac{1}{2}\angle A$ and the previous expression becomes
$R/\sin 45° = R\sin\frac{1}{2}\angle A / \sin\frac{1}{2}\angle C$,

or $\quad \sin\frac{1}{2}\angle A = \sqrt{2}\sin\frac{1}{2}\angle C \qquad$ (i)

We also have $\frac{1}{2}\angle A + \frac{1}{2}\angle C = 45°$ or $\sin(\frac{1}{2}\angle A + \frac{1}{2}\angle C) = \sin 45°$

Now expand it, $\sin\frac{1}{2}\angle A \cos\frac{1}{2}\angle C + \cos\frac{1}{2}\angle A \sin\frac{1}{2}\angle C = \sqrt{2}/2$

Substituting $\sin\frac{1}{2}\angle C$ from (i), we have

$\sin\frac{1}{2}\angle A \cos(45° - \frac{1}{2}\angle A) + \cos\frac{1}{2}\angle A \sin\frac{1}{2}\angle A\sqrt{2}/2 = \sqrt{2}/2$
or
$\sin\frac{1}{2}\angle A(\cos\frac{1}{2}\angle A + \sin\frac{1}{2}\angle A) + \cos\frac{1}{2}\angle A \sin\frac{1}{2}\angle A = 1$, or

$2\sin\frac{1}{2}\angle A \cos\frac{1}{2}\angle A = \cos^2\frac{1}{2}\angle A$, or $2\sin\frac{1}{2}\angle A = \cos\frac{1}{2}\angle A$, or

$\tan\frac{1}{2}\angle A = 0.5 = DI/AD$, or $DI = r$, $AD = 2r$ and $AI = r\sqrt{5}$, and
$\cos\angle A = \cos^2\frac{1}{2}\angle A - \sin^2\frac{1}{2}\angle A = (AD/AI)^2 - (DI/AI)^2 = 3/5 = .6$

But $\cos\angle A = AB/CA = 0.6$

Now applying the Pythagorean formula $CA^2 = AB^2 + BC^2$, we have

$CA^2 = 0.36\,CA^2 + BC^2$ or $BC/CA = 0.8$

Finally, $AB : BC : CA = 3 : 4 : 5$

Problem 2 of the British Mathematical Olympiad 2008

Let ABC be an acute-angled triangle with $\angle B = \angle C$. Let the circumcenter be O and the orthocenter be H. Prove that the center of the circle BOH lies on the line AB. The circumcenter of a triangle is the center of its circumcircle. The orthocenter of a triangle is the point where its three altitudes meet.

Solution

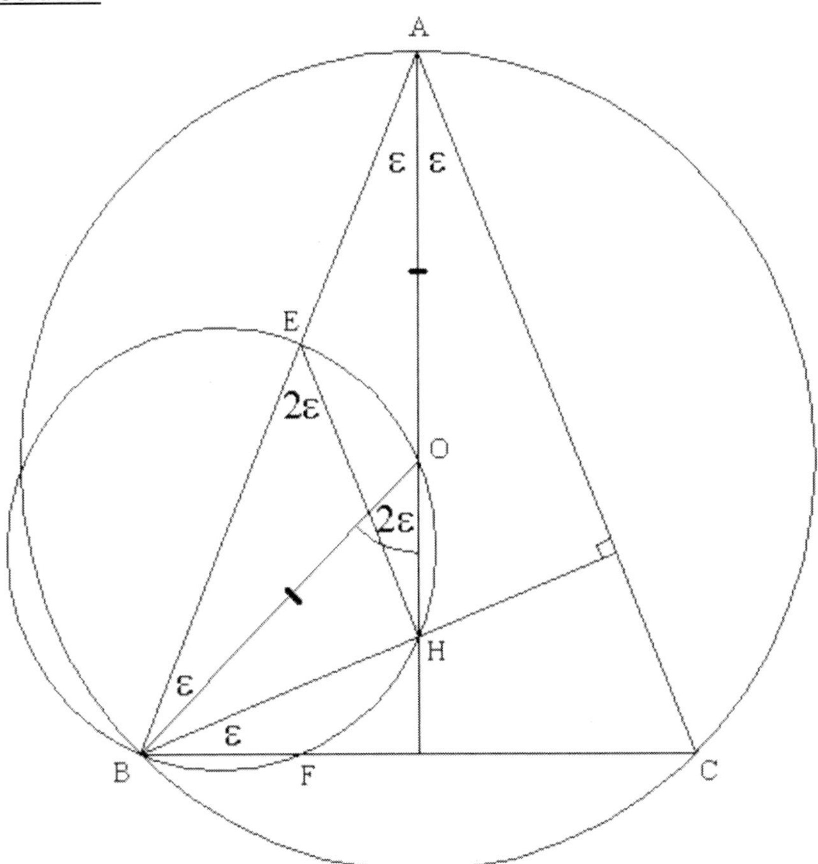

Let ε = /_HAB, we then also have ε = /_HAC = /_ABO (since O is center of circle), and /_BOH = /_BAO + /_ABO = 2ε.

Now let the BOH circle intercept AB at E.

We then have $\angle BEH = \angle BOH = 2\varepsilon$ (these angles subtend the same arc BH)

Since $\angle HBC + \angle ACB = 90° = \angle HAC + \angle ACB$
we have $\angle HBC = HAC = \varepsilon$

and $\angle HAB + \angle ABC = 90°$ or $\angle HAB + \angle ABO + \angle OBH + \angle HBC = 3\varepsilon + \angle HBO = 90° = \angle EBO + \angle HBO + \angle BEH$

or $\angle EBH + \angle BEH = 90°$ and $\angle BHE = 180° - \angle EBH - \angle BEH = 90°$, or

BE is the diameter of the circle BOH, or the center of the circle BOH lies on the line AB.

Problem 2 of the British Mathematical Olympiad 2009

In triangle ABC the centroid is G and D is the midpoint of CA. The line through G parallel to BC meets AB at E. Prove that $\angle AEC = \angle DGC$ if, and only if, $\angle ACB = 90°$. The centroid of a triangle is the intersection of the three medians, the lines which join each vertex to the midpoint of the opposite side.

Solution

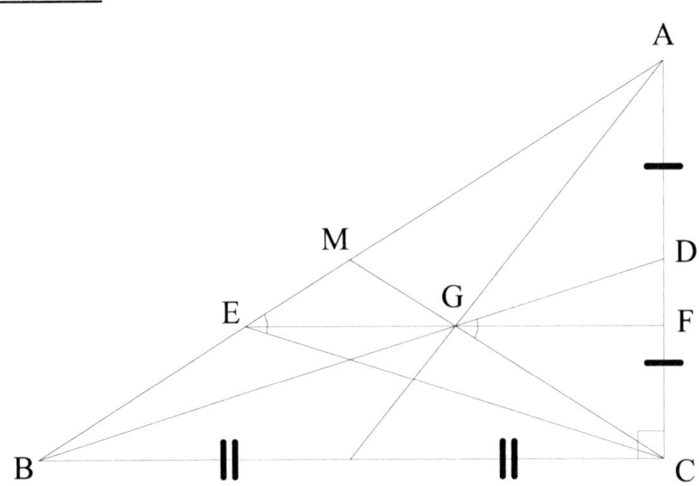

If $\angle ACB = 90°$

Let EG intercept AC at F. Since EF || BC, $\angle AEF = \angle ABC$, $\angle DGF = \angle DBC$, and $\angle FEC = \angle ECB$

We have $\angle AEC = \angle AEF + \angle FEC = \angle ABC + \angle ECB$

But since M is the midpoint of AB and $\angle ACB = 90°$, M is also the center of the circumcircle of triangle ABC and AM = MC = MB and $\angle ABC = \angle MCB$

Therefore, $\angle AEC = \angle ABC + \angle ECB = \angle MCB + \angle ECB$

But since EG || BC and MB = MC, we have $\angle$MBG = $\angle$MCE, or $\angle$ECB = $\angle$DBC.

or $\angle$AEC = $\angle$MCB + $\angle$ECB = $\angle$MCB + $\angle$DBC = $\angle$DGC,

<u>Now if $\angle$AEC = $\angle$DGC</u>

We have
$\angle$AEC = $\angle$ABC + $\angle$ECB = $\angle$EBG + $\angle$GBC + $\angle$ECB,

and
$\angle$DGC = $\angle$GBC + $\angle$GCB = $\angle$GBC + $\angle$GCE + $\angle$ECB,

or $\angle$EBG = $\angle$GCE or EGCB is cyclic.

Combining with EG || BC, we have EB = GC, and EGCB is an isosceles trapezoid, and $\angle$EBC = $\angle$GCB, or MBC is an isosceles triangle and MB = MC = MA.

Therefore,
$\angle$MBC = $\angle$MCB and $\angle$MAC = $\angle$MCA,

or $\angle$MCB + $\angle$MCA = ½ 180° = 90°.

Problem 3 of Asian Pacific Mathematical Olympiad 1989

Let A1, A2, A3 be three points in the plane, and for convenience, let A4 = A1, A5 = A2. For n = 1, 2, and 3, suppose that Bn is the midpoint of AnAn+1, and suppose that Cn is the midpoint of AnBn. Suppose that AnCn+1 and BnAn+2 meet at Dn, and that AnBn+1 and CnAn+2 meet at En. Calculate the ratio of the area of triangle D1D2D3 to the area of triangle E1E2E3.

Solution

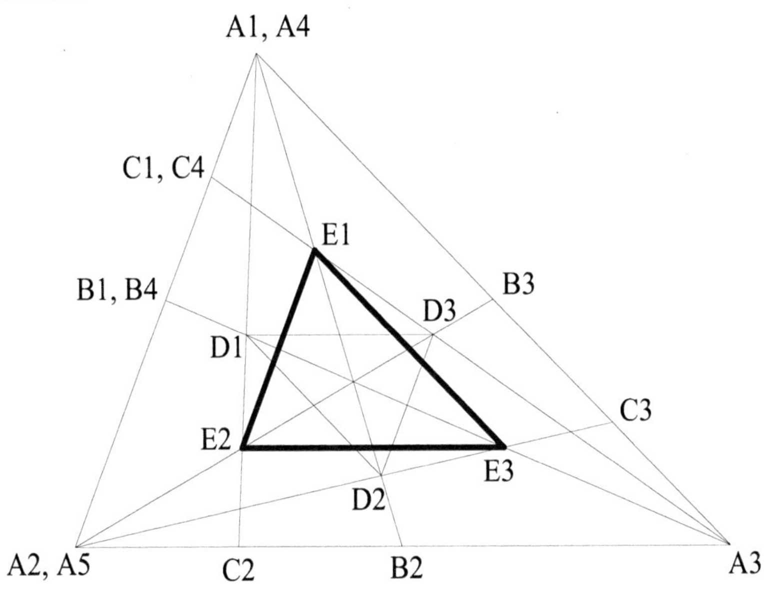

Connect and extend A1D3 to meet A2A3 at A, A2D1 to meet A1A3 at B and A3D2 to meet A1A2 at C.

Apply Ceva's theorem for the three lines A1C2, A3B1 and A2B, we have:
(A2C2 × A3B × A1B1) / (A3C2 ×A1B ×A2B1) = 1
since A3C2 = 3 ×A2C2 and A1B1 = A2B1,
we have A1B / A3B = 1/3

We then have A1C1 / A2C1 = 1/3.
Therefore, BC1 || A2A3, and C1B/A2A3 = A1C1/A1A2 = 1/4

We then have $(A2B2 \times A3B \times A1C1) / (A3B2 \times A1B \times A2C1)$
$= 1 \times 3 \times (1/3) = 1$
Therefore, per Ceva's theorem point E1 is on A2B.
With the same argument, E2 is on A3C and E3 is on A1A.

Now apply Ceva's theorem for the three lines E1B2, A2D3 and A3D1 that meet at G, we have
$E1D3 / A3D3 = E1D1 / A2D1$ or $D1D3 \parallel A2A3$
Similarly, $D2D3 \parallel A1A2$ and $D1D2 \parallel A1A3$.
and for the three lines GB2, A2E3 and A3E2 that meet at D2, we have $GE3 / A3E3 = GE2 / A2E2$ or $E2E3 \parallel A2A3$

Similarly, $E1E2 \parallel A1A2$ and $E1E3 \parallel A1A3$
$\Delta D1D2D3$ and $\Delta E1E2E3$ are similar since their corresponding sides are parallel to each other.

With the parallel lines, we now have:

$E1D1 / E1A2 = D1D3 / A2A3 = D1D3 / 2AC2 = A1D1 / 2A1C2 = A2D1 / 2A2B$, or

$A2D1 / 2E1D1 = A2B / E1A2 = (E1A2 + E1B) / E1A2 =$
$1 + E1B / E1A2 = 1 + C1B / A2A3 = 1 + 1/4 = 5/4$,
or $2E1D1 / A2D1 = 2E1D3 / A3D3 = 4/5$,
or $E1D3 / A3D3 = 2/5$

Adding 1 to both sides, we have $1 + E1D3 / A3D3 = 7/5$,

or $(A3D3 + E1D3) / A3D3 = 7/5$,
or $E1A3 / A3D3 = 7/5$,
or $A3D3 / E1A3 = 5/7$,
or $D2D3 / E2E1 = 5/7$

Thus the ratio of the corresponding sides of the two similar triangles $\Delta D1D2D3$ and $\Delta E1E2E3$ is equal to 5/7. Therefore the ratio of the area of $\Delta D1D2D3$ to the area of $\Delta E1E2E3$ is

$$(5/7) \times (5/7) = 25/49.$$

Problem 3 of Asian Pacific Mathematical Olympiad 1990

Consider all the triangles ABC which have a fixed base BC and whose altitude from A is a constant h. For which of these triangles is the product of its altitudes a maximum?

Solution

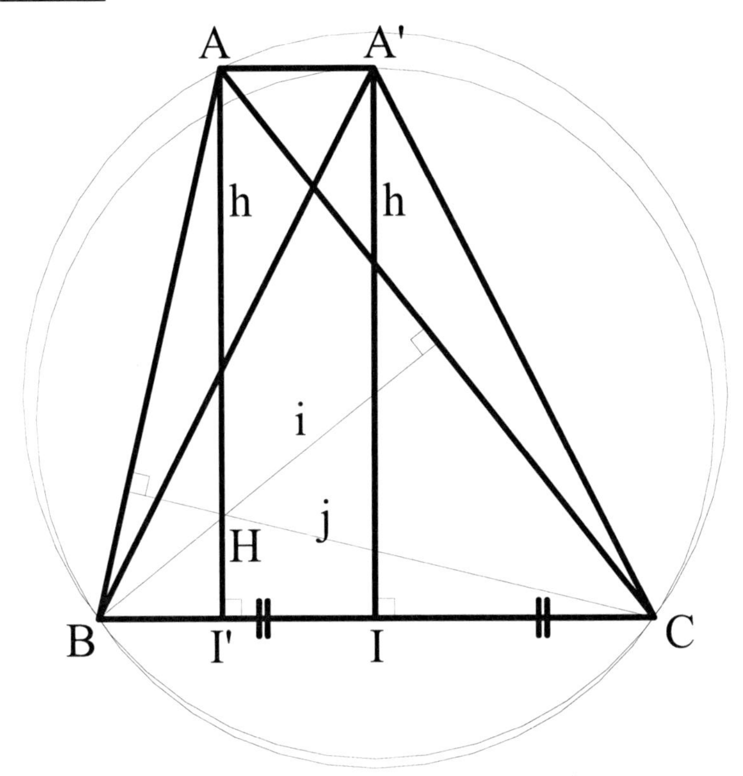

From B draw line perpendicular to AC and from C draw line perpendicular to AB. We have the altitudes i and j, respectively.

The problem asks for the product h times i times j to be maximum. But h is constant, so $i \times j$ must be a maximum.

The area of the triangle ABC is also constant since base BC is fixed.

Twice the area of triangle ABC = h × AC = i × AC = j × AB. From there,

i × AC × j × AB = square of twice the area of triangle ABC = constant

The multiplication of two products (i × j) and (AC × AB) is a constant, for one to be maximum (i × j) the other has to me minimum, we must find AC and AB so that

AB × AC = minimum.

Let R be the radius of the circumcircle of triangle ABC, and a, b and c as the lengths of its sides, there exists a formula

Area of ABC = abc / 4R

Area of ABC is fixed as we know, so for the product of the three sides to be a minimum (one side is already fixed, or the product of the two sides to be a minimum) the denominator R has to be minimum, or the circumcircle has to be smallest (A–> A') and A'B = A'C. The triangle is isosceles.

Problem 3 of Asian Pacific Mathematical Olympiad 1995

Let PQRS be a cyclic quadrilateral such that the segments PQ and RS are not parallel. Consider the set of circles through P and Q, and the set of circles through R and S. Determine the set I of points of tangency of circles in these two sets.

Solution

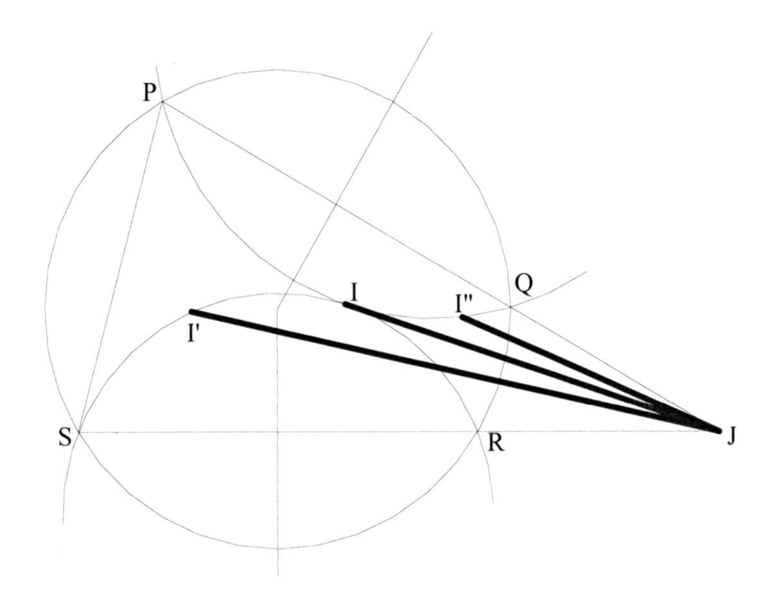

Extend PQ and SR to intercept each other at J. Since PQRS is a cyclic quadrilateral $\angle PSR + \angle PQR = 180°$ or $\angle PSR = \angle RQJ$

Similarly $\angle SPQ = \angle QRJ$

And the two triangles JPS and JRQ are similar; therefore,

$JQ / JR = JS / JP,$ or $JQ \times JP = JS \times JR$

From J draw the two lines tangential to the bottom and top circles and assume that the two tangential points are different, respectively, are I' and I" on the bottom and top circles as shown. We have

$JI'^2 = JR \times JS$ and $JI''^2 = JQ \times JP$

With the assumption that the tangential JI' and JI" are not coincided, the two circles are either overlap or not touching each other at all which is not true with the given condition of the problem. Therefore, for the two circles to tangent I' must coincide I" and also coincide with I.

or $JI^2 = JR \times JS$ which is a constant

So the set of points of tangency of the two circles is a circle with

center at J and radius $r = \sqrt{JR} \text{ x } JS$ or $r = \sqrt{JQ} \text{ x } JP$

Problem 3 of Asian Pacific Mathematical Olympiad 1999

Let *C1* and *C2* be two circles intersecting at *P* and *Q*. The common tangent, closer to *P*, of *C1* and *C2* touches *C1* at *A* and *C2* at *B*. The tangent of *C1* at *P* meets *C2* at *C*, which is different from *P*, and the extension of *AP* meets *BC* at *R*. Prove that the circumcircle of triangle *PQR* is tangent to *BP* and *BR*.

Solution

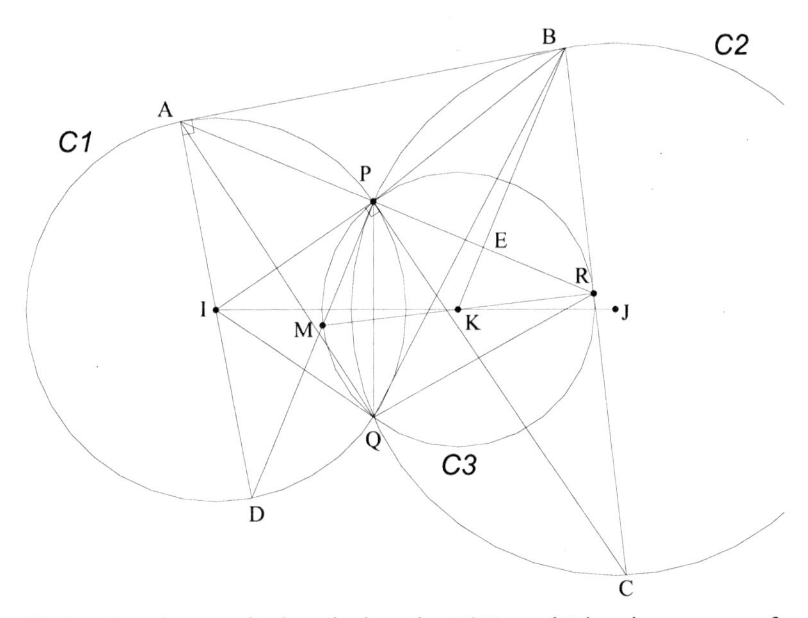

Let C3 be the circumcircle of triangle PQR and I be the center of circle *C1*.

We have $\angle RAQ + \angle IPQ = 90°$ (they combine to cut half the circle).

Therefore, $\angle RAQ = \angle QPC$ (i)

but $\angle QPC = \angle QBC$ and $\angle RAQ = \angle RBQ$ and A, B, R, Q are concyclic.

We have $\angle BPR = \angle BAR + \angle ABP$ (ii)

But $\angle BAR = \angle BQR$ (because of cyclic ABRQ) and $\angle ABP = \angle PQB$, and (ii) becomes

$\angle BPR = \angle PQB + \angle BQR = \angle PQR$ or BP is tangential to circle C3.

Now extend AI to intercept circle *C1* at D. Link PD to intercept *C3* at M

We have $\angle DAQ = \angle DPQ = \angle MRQ$ (iii)

We also have $\angle PAB + \angle API = 90°$ and $\angle API + \angle RPC = 180° - \angle IPC = 90°$

or $\angle PAB = \angle RPC$ (iv)

Combining (i) and (iv) we have $\angle QAB = \angle QPR$

But $\angle QAB + \angle DAQ = 90°$ therefore,

$\angle QPR + \angle DPQ = \angle MPR = 90°$, or

MR is the diameter of *C3.*

In cyclic quadrilateral ABRQ the angles
$\angle QAB + \angle QRB = 180°$ (v)

Adding $\angle DAQ$ and subtract $\angle MRQ$ from (iii) to the left side of (v), we have

$\angle QAB + \angle DAQ + \angle QRB - \angle MRQ = 180°$, or

$90° + \angle MRP = 180°$ or $\angle MRP = 90°$

Since MR is diameter of *C3* as proven earlier, therefore, BR is also tangential to circle C3.

Further observation

The problem below is derived from the above problem:

Let *C1* and *C2* be two circles intersecting at *P* and *Q*. The common tangent, closer to *P*, of *C1* and *C2* touches *C1* at *A* and *C2* at *B*. The tangent of *C1* at *P* meets *C2* at *C*, which is different from *P*, and the extension of *AP* meets *BC* at *R*. Let K be the center of circumcircle of triangle PQR. BK intercepts PR at E. Prove that E lies on the center with diameter AB.

Problem 3 of Asian Pacific Mathematical Olympiad 2000

Let *ABC* be a triangle. Let *M* and *N* be the points in which the median and the angle bisector, respectively, at *A* meet the side *BC*. Let *Q* and *P* be the points in which the perpendicular at *N* to *NA* meets *MA* and *BA*, respectively, and *O* the point in which the perpendicular at *P* to *BA* meets *AN* produced. Prove that *QO* is perpendicular to *BC*.

Solution

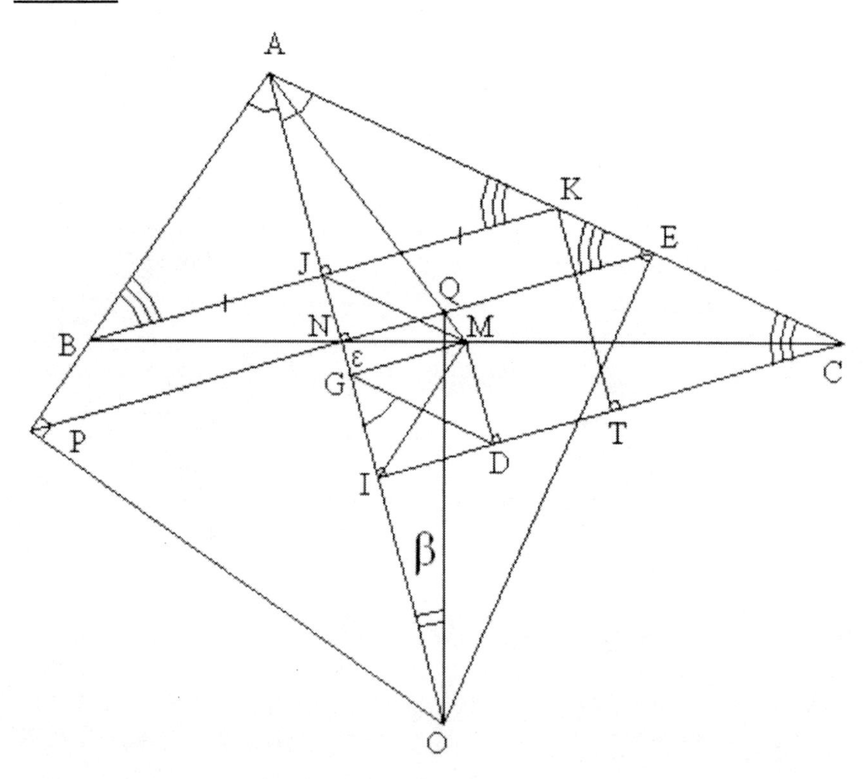

Let angle ONC = ε and angle NOQ = β

From B draw line || with PE to intercept AO at J and AC at K.
From C draw line || with PE to intercept AO at I.
From M draw line || with PE to intercept AO at G.
From M draw line || with AO to intercept IC at D.
From K draw line || with AO to intercept IC at T.

Because APOE is a cyclic quadrilateral and NP = NE, we have

$NO = NE^2 / NA$, and since $PE \parallel MG$, we have

$NQ / GM = NA / GA$, or $NQ = NA \times GM / GA$

Since M is the midpoint of BC we then have $GM = \frac{1}{2}(CI - BJ)$ and then

$NQ = \frac{1}{2} NA \times (CI - BJ) / GA$

Therefore, $\tan\beta = NQ / NO = \frac{1}{2} NA^2 (CI - BJ) / (GA \times NE^2) =$

$\frac{1}{2}(NA / NE)^2 \times (CI - BJ) / GA = \frac{1}{2}(IA / CI)^2 \times (CT / GA)$

To prove the two lines are perpendicular we need to prove

$\tan\beta \times \tan\varepsilon = 1$ but $\tan\varepsilon = CI / NI$ or we then need to prove

$\frac{1}{2}(IA / CI)^2 \times (CT / GA) \times (CI / NI) = 1$

or $(IA^2 / CI) \times CT = 2\, NI \times GA$ (i)

but $IA / CI = KT / CT$, (i) becomes

$IA \times KT = 2\, NI \times GA$ (ii)

But $NI = GN + GI$ $GI = GJ$ and $JI = KT$, equation (ii) becomes

$IA \times KT = (KT + 2\, GN) \times GA$

or $GI \times KT = 2\, GI^2 = 2\, GN \times GA$

or to prove $GN / GI = GI / GA$, but

$GN / GI = GN / GJ = NM / BM = NM / MC = ID / DC$

Now we need to prove that ID / DC = GI / GA

or $\quad GD \parallel AC$

Since J and M are midpoints of BK and BC, respectively, we have JM || AC;

therefore, $\quad \angle IJM = \angle IAC$

but GM is bisector of IJ, we then have JM = IM and $\angle IJM = \angle JIM$ or

$\angle JIM = \angle IAC$ but GMDI is a rectangle and we have $\angle IGD = \angle JIM$ or

$\angle IGD = \angle IAC$ or GD || AC, and the proof is complete.

Problem 3 of Asian Pacific Mathematical Olympiad 2002

Let *ABC* be an equilateral triangle. Let *P* be a point on the side *AC* and *Q* be a point on the side *AB* so that both triangles *ABP* and *ACQ* are acute. Let *R* be the orthocenter of triangle *ABP* and *S* be the orthocenter of triangle *ACQ*. Let *T* be the point common to the segments *BP* and *CQ*. Find all possible values of $\angle CBP$ and $\angle BCQ$ such that triangle *TRS* is equilateral.

Solution

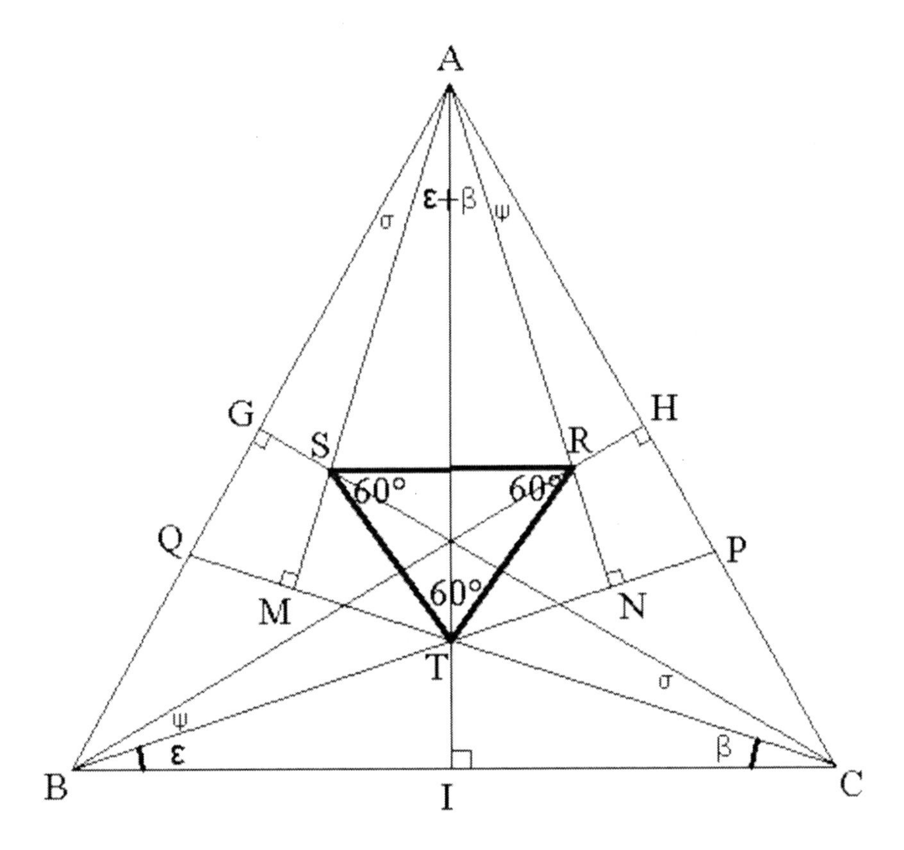

Let a be the side length of equilateral triangle ABC and

$$\angle CBP = \alpha$$
$$\angle BCQ = \beta$$
$$\angle SBR = \delta$$

$$\angle ABS = \sigma$$
$$\angle TBR = \psi$$

We have:

$BT / \sin\beta = a / \sin(\alpha+\beta)$ or $BT = a \sin\beta / \sin(\alpha+\beta)$

$BP / \sin 60° = PC / \sin\alpha = a / \sin(\alpha+60°)$

but $\alpha+60° = 90° - \psi$ therefore, $\sin(\alpha+60°) = \cos\psi$

$BP = \sin(60°)\, a/\cos\psi$ and $PC = a \sin\alpha / \cos\psi$

$AP = NP / \sin\psi$ or

$NP = \sin\psi\, AP = \sin\psi\, (a - PC) = a \sin\psi\, (1 - \sin\alpha / \cos\psi)$

$TN = BP - BT - NP$

$TN = \sin(60°)\, a / \cos\psi - a \sin\beta / \sin(\alpha+\beta) - a \sin\psi\, (1 - \sin\alpha/\cos\psi)$

Now for RN:

$RN / \sin\psi = BN / \cos\psi$ or $RN = BN \tan\psi = (BP - NP) \tan\psi$

$RN = [\sin(60°)\, a / \cos\psi - a \sin\psi\, (1 - \sin\alpha / \cos\psi)] \tan\psi$

$TR^2 = TN^2 + RN^2 = [\sin(60°)\, a/\cos\psi - a \sin\beta/\sin(\alpha+\beta) - a \sin\psi\, (1 - \sin\alpha / \cos\psi)]^2 + [\sin(60°) a / \cos\psi - a \sin\psi\, (1 - \sin\alpha / \cos\psi)]^2 \tan^2\psi$

Using the same process to find TS, we have

$TS^2 = TM^2 + SM^2 = [\sin(60°)\, a/\cos\sigma - a \sin\alpha/\sin(\alpha+\beta) - a \sin\sigma\, (1 - \sin\beta / \cos\sigma)]^2 + [\sin(60°)\, a / \cos\sigma - a \sin\sigma\, (1 - \sin\beta / \cos\sigma)]^2 \tan^2\sigma$

So for TR = TS one obvious solution is that $\alpha = \beta$, $\psi = \sigma$ to make the corresponding terms of TR^2 and TR^2 above equal, and when $\alpha = \beta$ the points P and Q are symmetrical across AI where I is the foot of A to BC.
Since SA = SB and CG $\perp$ AB and CQ $\perp$ AM, we then also have $\angle ABS = \angle SAB = \angle TCS = \sigma$
Since SA = SB and CG is perpendicular to AB and CQ perpendicular to AM, we then also have

$$\angle ABS = \angle SBA = \angle TCS = \sigma$$

Assume a solution has been attained and that $\angle CBP = \alpha 1$ and $\angle BCQ = \beta 1$ are the angles required for triangle TRS to be equilateral.

We will prove that for every unique value of angle $\alpha 1$ there is one and only one corresponding angle $\beta 1$ to satisfy the problem.

Indeed, let's keep angle $\alpha 1$ and increase $\angle BCQ$. As we do so point T moves to T' closer to N and RT' < RT, or RT decreases.

We also know that $\angle MAN = \alpha + \beta$. So $\angle MAN$ increase by the same amount of the increase of $\angle BCQ$, and $\angle GAS$ also decreases by the same amount. Therefore, as we increase $\angle BCQ$, point S moves to S' closer to point G and RS' > RS, or RS increases.

The same but opposite effect occurs if we decrease $\angle BCQ$.

Therefore, TR will no longer equal SR if $\angle BCQ \neq \beta 1$. So for every angle α there is only one unique angle β to satisfy the condition for triangle TRS to be equilateral.

We also know that $\angle CBP = \angle BCQ$ is a condition for ST = RT. So point T has to always be on AI, or $\alpha = \beta$. Now let's find $\angle \alpha$.

Since triangle TRS is equilateral, and R is on the bisector BH of $\angle ABC$, we have SR || BC, ST || AC and RT || AB, or BH is the bisector of $\angle SBT$ or $\delta = \psi$. We have

$$\sigma = 30° - \delta = 30° - \psi = \alpha$$

We also have $\quad \angle BCG = \sigma + \beta = 30°$

Therefore, $\quad \alpha = \beta = \delta = \sigma = \psi = 15°$

or $\quad \angle CBP = \angle BCQ = 15°$

Problem 3 of the Balkan Mathematical Olympiad 1988

Let ABCD be a tetrahedron and let d be the sum of squares of its edges' lengths. Prove that the tetrahedron can be included in a region bounded by two parallel planes, the distances between the planes being at most $\frac{1}{2}\sqrt{d/3}$.

Solution

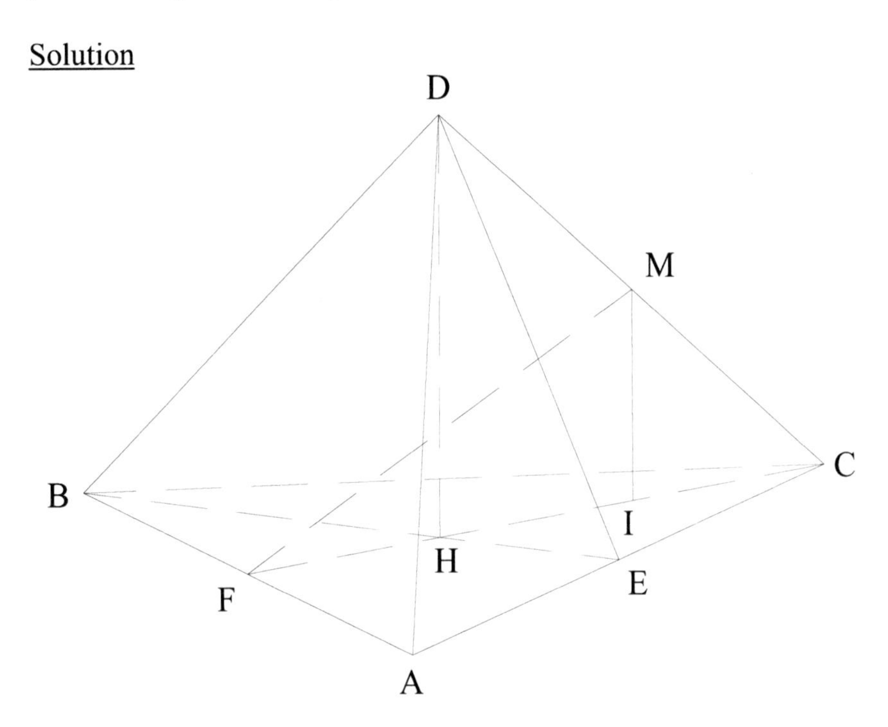

Let E, F and M be the midpoints of AC, AB and DC, respectively; also let the edge's length of the tetrahedron be l.

From D and M draw the altitudes to plane ABC to meet it at H and I, respectively. We will prove that the tetrahedron fits into the parallel planes with DC and AB on either plane.

The sum of squares of six lengths is $6l^2 = d$ or $l = \sqrt{d/6}$

Consider equilateral triangle DAC, $DE^2 = l^2 - l^2/4$, or $DE = l\sqrt{3}/2$.

We also have BE = DE and since H is also the centroid of triangle ABC,

$HE = BE/3 = DE/3 = l\sqrt{3}/6.$

Now consider right triangle DHE where $DH^2 = DE^2 - HE^2$

$= 3l^2/4 - 3l^2/36 = 2l^2/3$ or $DH = l\sqrt{2/3} = \sqrt{d/6}\sqrt{2/3} = \sqrt{d}/3$

Now $FM^2 = MI^2 + FI^2 = (DH/2)^2 + (2HE)^2 = \frac{1}{2}\sqrt{d/3}$, but as we can see FM is orthogonal to AB (in triangle BMA) and it's also orthogonal to DC (in triangle DFC).

Therefore, the plane containing AB and the plane containing DC that are both orthogonal to FM are parallel to each other. The tetrahedron, therefore, fits into the two planes being at most $\frac{1}{2}$

$\sqrt{d/3}$ apart.

Problem 3 of Belarusian Mathematical Olympiad 1997

Points D,M,N are chosen on the sides AC,AB,BC of a triangle ABC respectively, so that the intersection point P of AN and CM lies on BD. Prove that BD is a median of the triangle if and only if $\frac{AP}{PN}=\frac{CP}{PM}$.

Solution

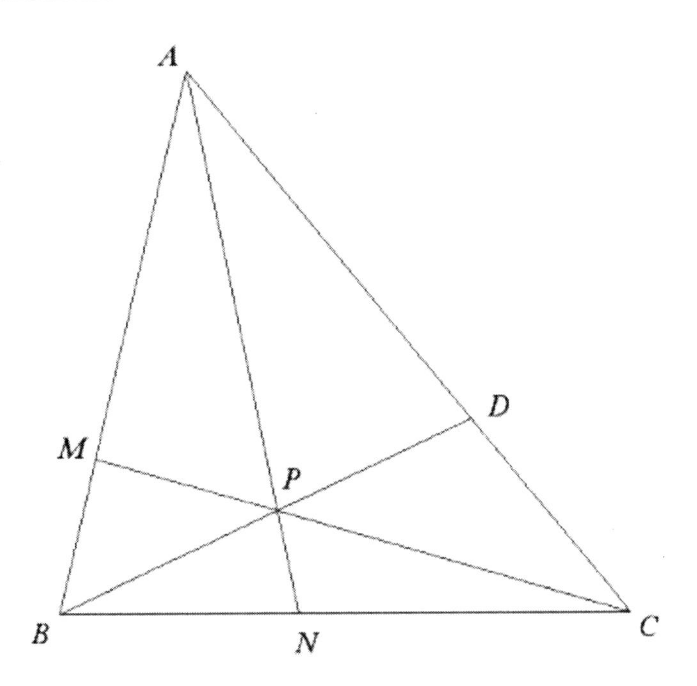

a) When BD is the median or AD = DC

Applying Ceva's theorem for the three concurrent segments AN, BD and CM, we have

$$\frac{BN}{NC}\times\frac{CD}{DA}\times\frac{AM}{MB}=1 \qquad \text{(i)}$$

and with AD = DC, we have $\frac{BN}{NC}\times\frac{AM}{MB}=1$ or

$$\frac{AM}{MB} = \frac{NC}{BN} \qquad \text{(ii)}$$

With AD = DC, we also have $\frac{AD}{DC} = \frac{DC}{AD}$

Now adding $\frac{AD}{DC}$ to the left of (ii) and $\frac{DC}{AD}$ to its right, we then have $\frac{AM}{MB} + \frac{AD}{DC} = \frac{NC}{BN} + \frac{DC}{AD}$

According to Van Aubel's theorem , we have

$\frac{AP}{PN} = \frac{AM}{MB} + \frac{AD}{DC}$ and $\frac{CP}{PM} = \frac{NC}{BN} + \frac{DC}{AD}$

Therefore, $\frac{AP}{PN} = \frac{CP}{PM}$

b) *When* $\frac{AP}{PN} = \frac{CP}{PM}$

Again, Van Aubel's theorem gives us $\frac{AP}{PN} = \frac{AM}{MB} + \frac{AD}{DC}$ and $\frac{CP}{PM} = \frac{NC}{BN} + \frac{DC}{AD}$

So when $\frac{AP}{PN} = \frac{CP}{PM}$, we have $\frac{AM}{MB} + \frac{AD}{DC} = \frac{DC}{AD} + \frac{NC}{BN}$ (iii)

Now let $x = \frac{AM}{MB}$, $y = \frac{AD}{DC}$, $z = \frac{NC}{BN}$

Equation (i) becomes $x \times \frac{1}{y} \times \frac{1}{z} = 1$

Equation (iii) becomes $x + y = \frac{1}{y} + z$

Those two equations give us $(z + 1)\,x^2 - z^2\,x - z^2 = 0$
Which has the single acceptable solution $x = z$, or

$\frac{AM}{MB} = \frac{NC}{BN}$ and from (iii), $\frac{AD}{DC} = \frac{DC}{AD}$ or AD = DC, or BD is a median of the triangle.

Problem 3 of Canadian Mathematical Olympiad 1986

A chord ST of constant length slides around a semicircle with diameter AB. M is the mid-point of ST and P is the foot of the perpendicular from S to AB. Prove that angle SPM is constant for all positions of ST.

Solution

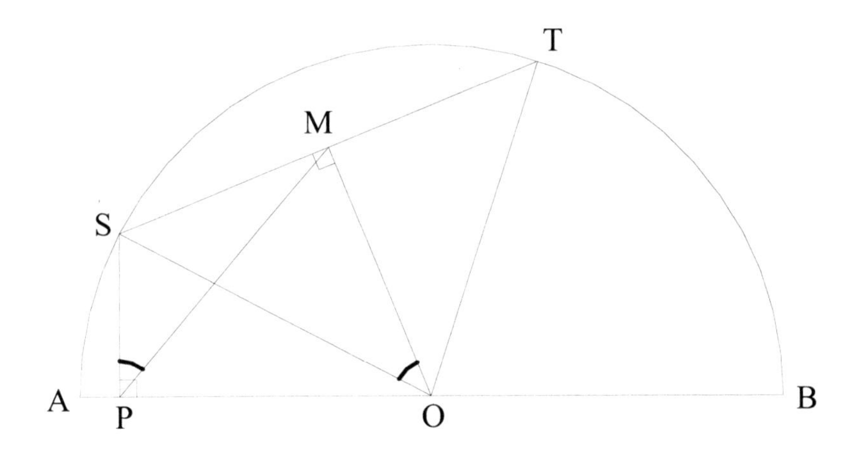

As the chord ST slides around AB, we note that triangle SOT rotates around O, and its shape remains constant, so is angle $\angle SOM$. Since M is the midpoint of ST, SM = MT.

We also have OS = OT = radius of the semicircle. The two triangles SOM and TOM are congruent because of their three sides are equal. Therefore, $\angle SMO = \angle TMO = \frac{1}{2}\, 180° = 90°$.

So the quadrilateral SPOM is cyclic since $\angle SPO + \angle SMO = 180°$. Therefore, $\angle SPM = \angle SOM$ is constant.

Problem 3 of Canadian Mathematical Olympiad 1991

Let C be a circle and P a given point in the plane. Each line through P which intersects C determines a chord of C. Show that the midpoints of these chords lie on a circle.

Solution

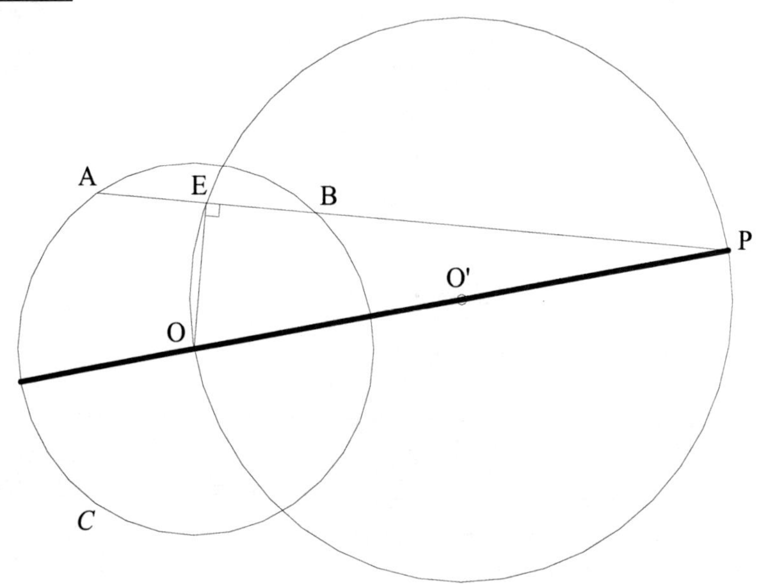

It's easily recognized that such a circle will pass through the center O of the circle C.

From P draw chord AB on C. Let E be its midpoint. Since OA = OB = radius OE $\perp$ AP so the two points O and E will be on the circumcircle of triangle OEP.

The above process applies to any chords initiating from P and intersecting circle C. Therefore, all the midpoints lie on the circumcircle of triangle OEP and it has the center at the midpoint of OP.

Problem 3 of the Canadian Mathematical Olympiad 1992

In the diagram, ABCD is a square, with U and V interior points of the sides AB and CD respectively. Determine all the possible ways of selecting U and V so as to maximize the area of the quadrilateral PUQV.

Solution

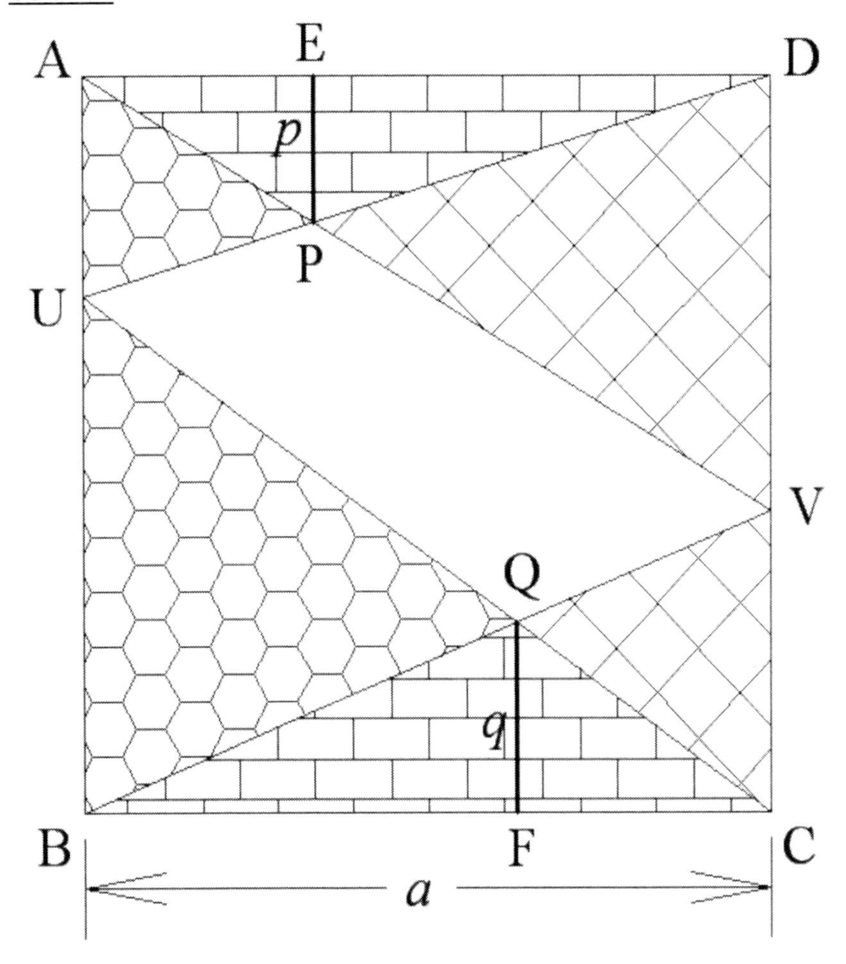

Let the side of the square be *a*. From P and Q draw perpendiculars to AD and BC, respectively, and let

PE = p and QF = q. Let's also denote (Ω) the area of shape Ω.

Note that the area of the quadrilateral PUQV is maximum when the total of the shaded areas is minimum.

It's easily seen that the total areas shaded with honey and bricks (AUD) + (BUC) = $\frac{1}{2}a$ (AU + UB) = $\frac{1}{2}a^2$ and is constant. So now the total areas shaded with squares (PDV) + (QVC) must be minimum.

But also note that (PDV) + (QVC) = (ADV) + (BCV) – (APD) – (BQC) = $\frac{1}{2}a^2$ – (APD) – (BQC)
so (PDV) + (QVC) is minimum when (APD) + (BQC) is maximum.

(APD) + (BQC) = ½ a(p + q) so the requirement now is for p + q to be a maximum.
Since both EP and QF || with the vertical sides of the square, we have

$$\frac{p}{AU} = \frac{DE}{a} = \frac{a - AE}{a} = 1 - \frac{AE}{a} = 1 - \frac{p}{DV}, \text{ or } \qquad p \times \frac{AU + DV}{AU \times DV} = 1,$$

or
$$p = \frac{AU \times DV}{AU + DV}$$

Similarly,
$$q = \frac{BU \times VC}{BU + VC}$$

$$p + q = \frac{AU \times DV}{AU + DV} + \frac{BU \times VC}{BU + VC} =$$

$$\frac{AU \times DV \times BU + AU \times DV \times VC + AU \times BU \times VC + BU \times VC \times DV}{AU \times BU + AU \times VC + DV \times BU + DV \times VC}$$

$$= \frac{AU \times BU\,(DV + VC) + DV \times VC\,(AU + BU)}{AU \times BU + AU \times VC + DV \times BU + DV \times VC} = a \times$$

$$\frac{AU \times BU + DV \times VC}{AU \times BU + AU \times VC + DV \times BU + DV \times VC}$$

Now divide both numerator and denominator by sum of products AU × BU + DV × VC, we have

$$p + q = a / \left(1 + \frac{AU \times VC + DV \times BU}{AU \times BU + DV \times VC}\right)$$

so now for p + q to be maximum, $\frac{AU \times VC + DV \times BU}{AU \times BU + DV \times VC}$ has to be a minimum. Let it be k.

But AU = a – BU and DV = a – VC, and

$$k = \frac{AU \times VC + DV \times BU}{AU \times BU + DV \times VC} \text{ becomes}$$

$$k = \frac{(a - BU)\,VC + (a - VC)\,BU}{(a - BU)\,BU + (a - VC)\,VC}$$

$$= \frac{a\,(VC + BU) - 2\,VC \times BU}{a\,(VC + BU) - (VC^2 + BU^2)}$$

$$= \frac{a\,(VC + BU) - 2\,VC \times BU}{a\,(VC + BU) - 2\,VC \times BU - (VC - BU)^2}$$

$$= 1 / \left[1 - \frac{(VC - BU)^2}{a\,(VC + BU) - 2\,VC \times BU}\right]$$

for k to be minimum the denominator of

$\frac{(VC - BU)^2}{a\,(VC + BU) - 2\,VC \times BU}$ has to be a maximum and

$\frac{(VC - BU)^2}{a\,(VC + BU) - 2\,VC \times BU}$ to be minimum. Note that the denominator is not zero, and the square $(VC - BU)^2$ is always greater than or equal to zero, and it's a minimum when it's zero or when VC = BU.

So to maximize the area of the quadrilateral PUQV, U and V has to be on a horizontal line between the top and bottom sides of the square ABCD. The maximal area of PUQV is then equal

$$a^2 - \tfrac{1}{2} a^2 - \tfrac{1}{2} (a/2) \times a = \tfrac{1}{4} a^2.$$

Problem 2 of the Ibero-American Mathematical Olympiad 1985

Let P be a point in the interior of the equilateral triangle ABC such that PA = 5, PB = 7, PC = 8. Find the length of the side of the triangle ABC.

Solution

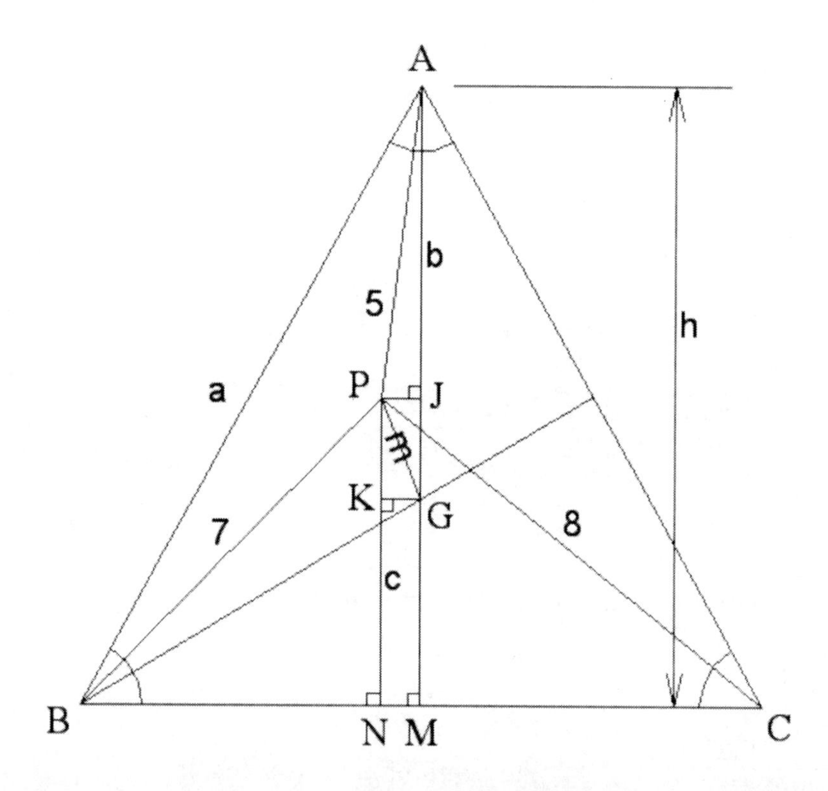

We know that there is an existing formula relating the distances from a point inside a triangle to its vertices expressed as follows $a^2 + b^2 + c^2 = 3\,(d^2 + e^2 + f^2 - 3m^2)$ where a, b and c are the lengths of the sides of the triangle; d, e and f are the distances from that point to the vertices, and m is the distance from that point to the triangle's centroid.
(http://www.gogeometry.com/problem/p255_triangle_centroid_squares_point_vertex.htm)

Let a be the side length of the equilateral triangle; apply the above formula we have

$$a^2 = d^2 + e^2 + f^2 - 3m^2$$

In our case let d = 5, b = 7 and f = 8 as given by the problem, we have $\quad a^2 = 138 - 3m^2 \quad$ (i)

now let's find m = PG as shown on the graph. Let AJ = b, we have

$$m^2 = PJ^2 + PK^2 = 25 - b^2 + (2/3\ h - b)^2 = 25 + 4/9\ h^2 - 4/3\ hb$$

where $h = a\sqrt{3}/2$ is the equilateral triangle's altitude

Substituting m^2 into (i), we have $a^2 = 63 - 4/3\ h^2 + 4\ hb \quad$ (ii)
Now substituting h into (ii), we have

$$2a^2 = 63 + 2ab\sqrt{3} \quad \text{or} \quad b = (2a^2 - 63)/[2a\sqrt{3}\,] \qquad \text{(iii)}$$

Now let s be the semi-perimeter of triangle PBC, s = (a + 15)/2, and the area of triangle PBC using Heron's formula gives us

$$\text{Area of triangle PBC} = \sqrt{(a + 15)(a + 1)(a - 1)(15 - a)} = 2a(h - b)$$

or $\quad (a^2 - 225)(a^2 - 1) = -4a^2 [a\sqrt{3}/2 - b]^2 \quad$ (iv)

Substituting b from (iii) into (iv), we have

$$(a^2 - 225)(a^2 - 1) = -4a^2 \{a\sqrt{3}/2 - 63/(2a\sqrt{3}\,)\}^2$$

Let $x = a^2$; the above equation reduces to a quadratic equation
$x^2 - 138x + 1161 = 0$ or $x^2 = 9$ and $x^2 = 129$
Therefore, a = 3 and a = 11.36 (11.35781669)

Length a can not be less than 5; we then pick 11.36 as the answer.

Problem 3 of the Ibero-American Mathematical Olympiad 1992

In a equilateral triangle of length 2, it is inscribed a circumference *C*.
(a) Show that for all point P of *C* the sum of the squares of the distance of the vertices A, B and C is 5.
(b) Show that for all point P of *C* it is possible to construct a triangle such that its sides has the length of the segments AP, BP and CP, and its area is $\frac{1}{4}\sqrt{3}$.

Solution

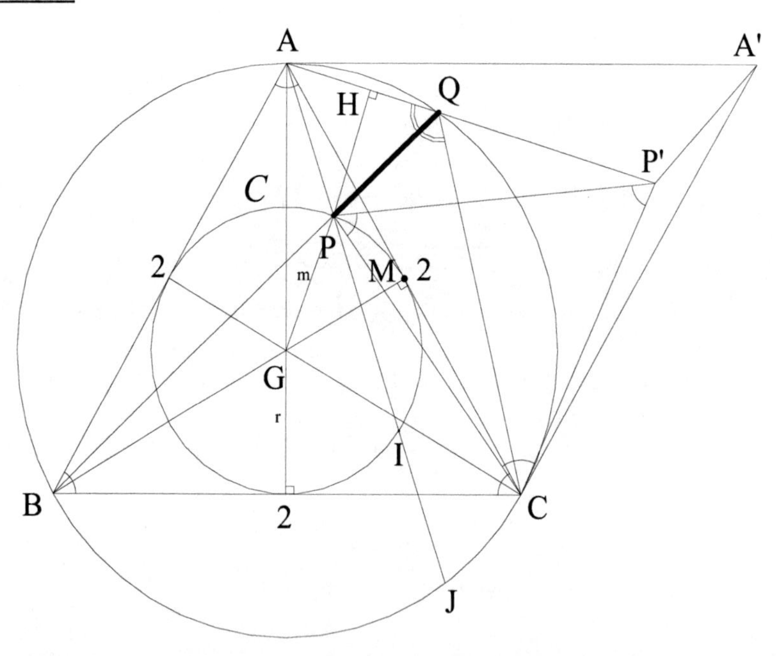

(a) We have the existing formula relating the distances from a point to the vertices of a triangle $a^2 + b^2 + c^2 = 3(d^2 + e^2 + f^2 - 3m^2)$ where a, b and c are the three sides of the triangle and d, e and f are distances from the point to the triangle's vertices. Applying the formula to this case with $m = GP = r = 1/\sqrt{3}$,
$4 = AP^2 + BP^2 + CP^2 - 3/3$ or $AP^2 + BP^2 + CP^2 = 5$

(b) Rotate triangle ABC 60° clockwise around point C. We have A –> A', B –> A, P –> P' and AP' = BP, PC = PP'=P'C and the triangle AP'P has its side lengths of the segments AP, BP and CP.

Now draw triangle ABC's circumcircle, and let Q be the intersection of AP' with the circumcircle. Since after the rotation, triangle ABP = triangle A'AP', $\angle$ABP = $\angle$A'AP', the three points B, P and Q are collinear. Let H be the foot of P to AP', the area of the triangle APP' = ½ PH × AP'. Now extend AP to intercept the two circles at I and J, respectively. We have PQ × PB = AP × PJ, but since the two circles share the same centers, AP = IJ and PQ × PB = AP × AI = AM^2 = 1

However, PB = AP' and $\angle$AQB = $\angle$BQC = 60° and PH = ½ $PQ\sqrt{3}$, the area of the triangle APP' = ½ PH × AP' = (¼ $PQ\sqrt{3}$) × PB = ¼ $\sqrt{3}$ PQ × BP = ¼ $\sqrt{3}$.

Further observation

The following are drawn from this problem:

1. Sum of distances from point Q to vertices of triangle ABC is

$AQ^2 + BQ^2 + CQ^2 = 8$ (i) since GQ = 2r = $2/\sqrt{3}$ or $BQ^2 = 8 - AQ^2 - CQ^2$.

2. Using the law of the cosine function $AC^2 = AQ^2 + CQ^2 - 2\,AQ \times CQ \cos 120°$ or $AQ^2 + CQ^2 + AQ \times CQ = 4$ (ii) or $(AQ + CQ)^2 = AQ \times CQ + 4$ (iii)
Now subtract (ii) from (i), we have $BQ^2 = AQ \times CQ + 4$
Combining with (iii), we have BQ = AQ + CQ.

3. The area of triangle with length segments AP, BP and CP is always constant as long as P is on the inner circle. One can

derive another problem to find the locus of the points P in the plane of an equilateral triangle ABC for which the triangle formed with PA, PB and PC has constant area.

4. The problem can be reversed: If for every point P in the interior of a triangle, one can construct a triangle having sides equal to PA, PB and PC then the triangle is equilateral.

5. In triangle ABC, AB is the longest side. Prove that for any point P in the interior of the triangle, PA + PB > PC.

Problem 3 of the Ibero-American Mathematical Olympiad 2002

Let P be a point in the interior of the equilateral triangle ABC such that $\angle APC = 120°$. Let M be the intersection of CP with AB, and N the intersection of AP and BC. Find the locus of the circumcenter of the triangle MBN when P varies.

Solution

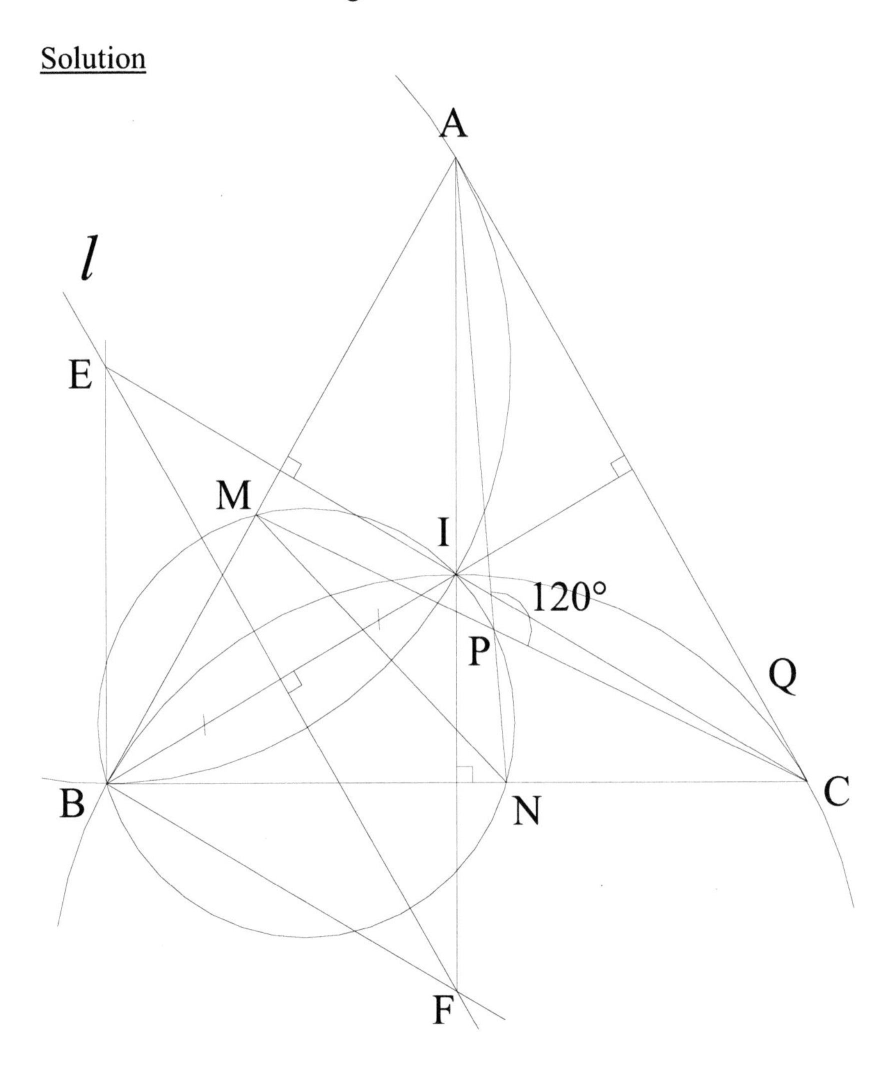

Since $\angle MPN = 120°$ and ABC is an equilateral triangle and $\angle ABC = 60°$, BMPN is cyclic.

We also noted that the circumcircle of triangle MBN has to pass through I, the incenter/ circumcenter/ centroid/ orthocenter of triangle ABC.

So, the circumcenter of triangle MBN passes through two fixed points B and I.

Thus the locus is on line l, the bisector of BI and $l \parallel AC$. The locus is from E to F excluding points E and F where EF = AC, the length of the triangle ABC since beyond those two points the circles do not cut the side of triangle ABC.

Problem 3 of the International Mathematical Olympiad 1960

In a given right triangle ABC; the hypotenuse BC, of length a, is divided into n equal parts (n an odd integer). Let α be the acute angle subtending, from A; that segment which contains the midpoint of the hypotenuse. Let h be the length of the altitude to the hypotenuse of the triangle. Prove:

$$\tan \alpha = \frac{4nh}{(n^2 - 1)\, a}$$

Solution

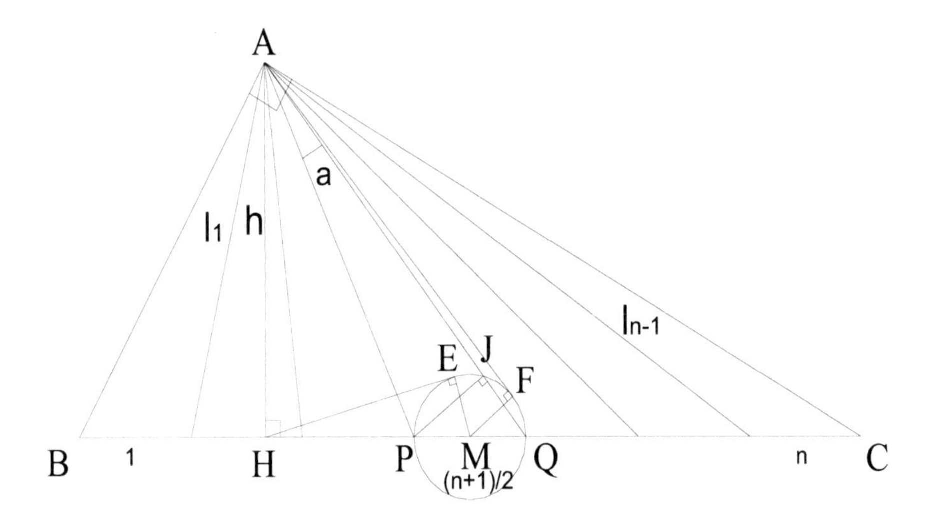

Let the two left and right segments meeting at A to make up angle α meet BC at P and Q, respectively. We have PQ = a/n. Let H be the foot of A to BC, and let AH be h.

From P draw a line to perpendicular to and meet AQ at J. Draw a circle with radius PQ and let its radius be called r. From A draw the tangent to the circle and meet it at F; from H draw another tangent to the circle and meet it at E.

We have $\tan\alpha = \dfrac{PJ}{AJ} = \dfrac{PJ \times AQ}{AJ \times AQ} = \dfrac{PJ \times AQ}{AF^2} = \dfrac{PJ \times AQ}{AM^2 - r^2} =$

$$\frac{PJ \times AQ}{AH^2 + HM^2 - r^2} = \frac{PJ \times AQ}{AH^2 + HE^2} = \frac{PJ \times AQ}{AH^2 + HP \times HQ}$$

But PJ × AQ is twice the area of APQ, and it is also equal AH × PQ; we then have

$$\tan\alpha = \frac{AH \times PQ}{AH^2 + HP \times HQ} = \frac{AH \times PQ}{AH^2 + (MH - a/2n)(MH + a/2n)}$$

$$= \frac{ha}{n\,[AH^2 + MH^2 - (a^2/4n^2)]} = \frac{ha}{n\,[AM^2 - (a^2/4n^2)]}$$

$$= \frac{ha}{n\,[(a^2/4) - (a^2/4n^2)]} = \frac{4nh}{(n^2 - 1)\,a}$$

Problem 3 of the Irish Mathematical Olympiad 2006

Prove that a square of side 2.1 units can be completely covered by seven squares of side 1 unit.

Solution

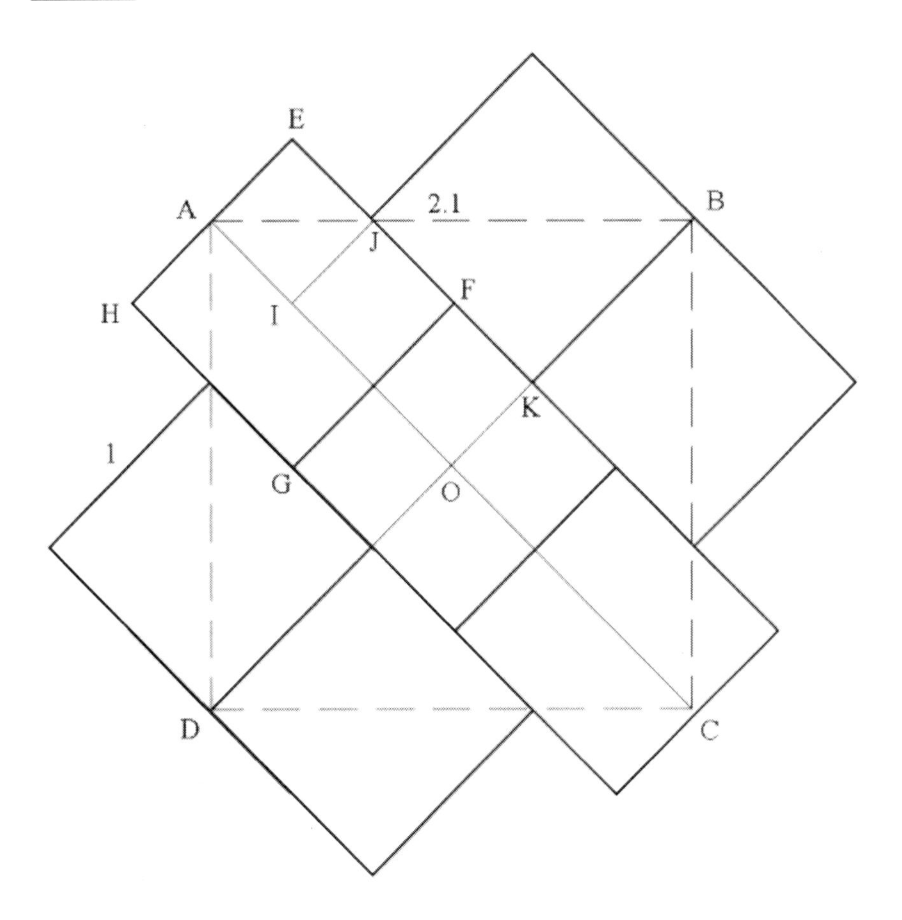

Let the square of side 2.1 units be ABCD and one of the seven equaled squares of side 1 unit being EFGH. Lay EFGH with its sides EF and HG to parallel with the diagonal AC of ABCD and the midpoint of EH to coincide with A as shown. Lay the rest of

the smaller squares side by side starting with side FG such that they are all adjacent to one another. Let O be the intersection of the two diagonal lines of ABCD. Now let the sides of the smaller squares intercept AB and BD at J and K, respectively.

We have

$AC = AB\sqrt{2} = 2.1\sqrt{2} = 2.97 < 3 \times 1 =$ sum of the three sides of the three smaller squares. So the sum of the three sides of the smaller squares is greater than the diagonal of ABCD.

It's easily seen that

$AI = ½\, EF = 0.5 = JI$ and $AJ = AI\sqrt{2} = 0.5\sqrt{2}$

We now have $JK/AO = BJ/AB$ or

$JK = AO \times BJ/AB = (AC/2) \times (AB–AJ)/AB = (2.1\sqrt{2}/2) x (2.1 -$

$0.5\sqrt{2})/2.1 = 0.9849$

which is smaller than 1, the unit of the side of the smaller square. Therefore, seven smaller squares of side 1 unit can cover a square of side 2.1 units the way we lay them out.

Problem 4 of the Austrian Mathematical Olympiad 2008

In a triangle ABC let E be the midpoint of the sides AC and F the midpoint of the side BC. Furthermore let G be the foot of the altitude through C on the side AB (or its extension). Show that the triangle EFG is isosceles if and only if ABC is isosceles.

Solution

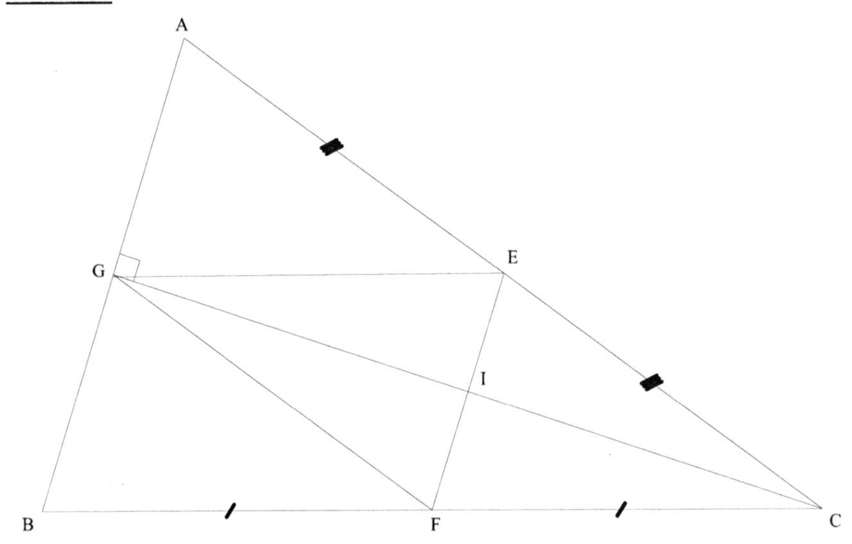

Let CG intercept EF at I. We only solve the problem with one geometrical configuration. Other configurations can also be solved similarly.

First assume that the triangle EFG is isosceles and GE = GF, $\angle GEF = \angle GFE$.
Since E and F are the midpoints of AC and BC, respectively, EF || AB and $\angle GEF = \angle AGE$ and $\angle GFE = \angle BGF$. Combining with $CG \perp AB$, we have

$\angle EGC = \angle FGC$. The two triangles EGC and FGC are then congruent since they also share GC. It follows that EC = FC and AC = BC or triangle ABC is isosceles.

Now assume triangle ABC is isosceles, AC = BC and EC = FC, $\angle BAC = \angle ABC$.

Since $\angle AGC = \angle BGC = 90°$, two triangles AGC and BGC are congruent and AG = BG which leads us to $\angle ACG = \angle BCG$. The two triangles EGC and FGC are then congruent since they also share GC.

It follows that EG = FG and triangle EFG is also isosceles.

Problem 4 of the Austrian Mathematical Olympiad 2009

Let D, E and F be the midpoints of the sides of the triangle ABC (D on BC, E on CA and F on AB). Further let HaHbHc be the triangle formed by the base points of the altitudes of the triangle ABC. Let P, Q and R be the midpoints of the sides of the triangle HaHbHc (P on HbHc, Q on HcHa and R on HaHb).

Show: The lines PD, QE and RF share a common point.

Solution

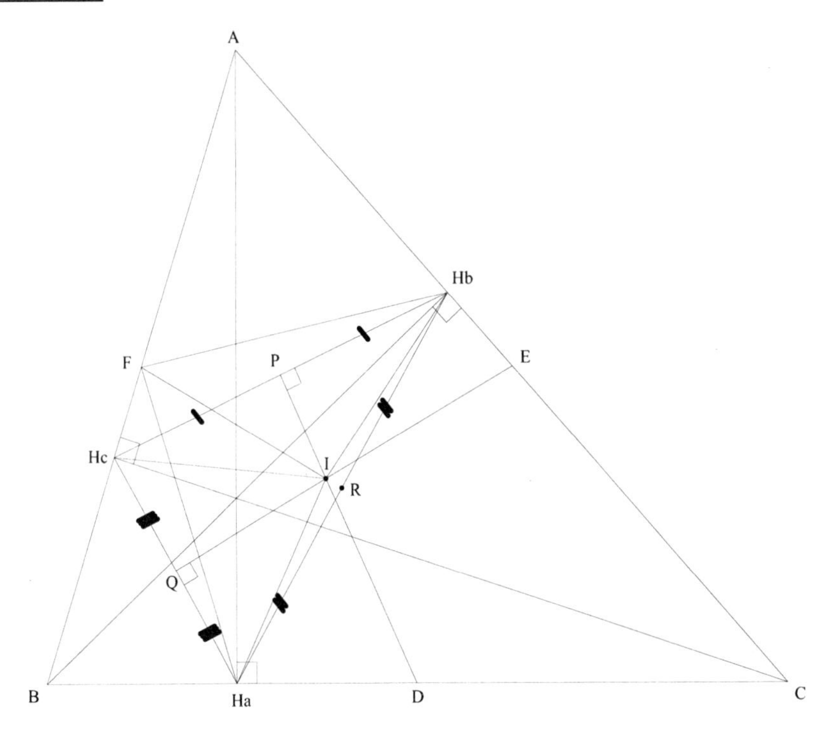

Let I be the intersection of DP and EQ. Since BHcC and BHbC are right triangles,

BHcHbC is cyclic, and with D being the midpoint of diameter BC, DHc = DHb.
Combining with P being the midpoint of HbHc, we have DP ⊥ HbHc.

Similarly, AHcHaC is cyclic, EHc = EHa and EQ ⊥ HaHc. Also FR ⊥ HaHb.

Since I is on DP and EQ and DP ⊥ HbHc, EQ ⊥ HaHc, we have IHb = IHc and IHa = IHc or IHa = IHb or IR ⊥ HaHb since R is also the midpoint of HaHb.

Combining with FR ⊥ HaHb, the three points F, I and R are collinear, or the lines PD, QE and RF share a common point I.

Problem 4 of Asian Pacific Mathematical Olympiad 1995

Let *C* be a circle with radius *R* and center *O*, and *S* a fixed point in the interior of *C*. Let *AA'* and *BB'* be perpendicular chords through *S*. Consider the rectangles *SAMB*, *SBN'A'*, *SA'M'B'*, and *SB'NA*. Find the set of all points *M*, *N'*, *M'*, and *N* when *A* moves around the whole circle.

Solution

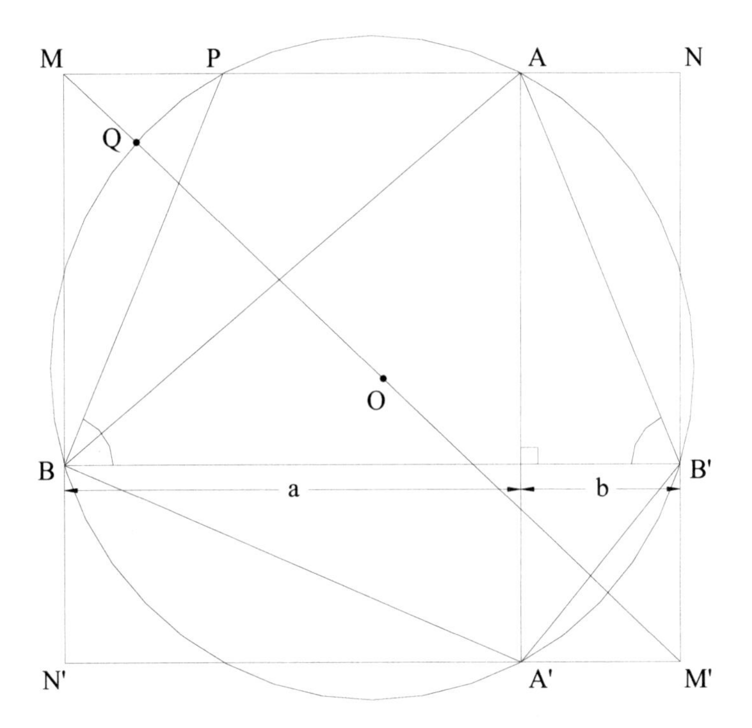

Let r be the radius of the circle, a = SB and b = SB'. Also let MA and MO intercept the circle at P and Q, respectively.

Now let MQ = c.

Since MA || BA, we have BP = B'A.

$\angle PBB' = \angle AB'B$, or $\angle PBM = \angle AB'N$

triangle PBM = triangle AB'N; therefore, MP = NA = SB' = b

From point M outside the circle, we have:

$MP \times MA = MQ \times (MQ + 2r)$

or $\quad a \times b = c(c + 2r)$

or $\quad c^2 + 2rc - ab = 0$

we have $c = -r \pm \sqrt{R^2 + ab}$

Therefore $\quad OM = c + r = \sqrt{R^2 + ab}$

The same proof can be used for other points N', M' and N; we have

$OM = ON' = OM' = ON = \sqrt{R^2 + ab}$

Since S is a fixed point inside circle c, the product ab is fixed. From there we conclude that the set of all points *M, N', M'*, and *N* when *A* moves around the whole circle is a circle with

the radius $R = OM = \sqrt{R^2 + ab}$

Problem 4 of Asian Pacific Mathematical Olympiad 1998

Let *ABC* be a triangle and *D* the foot of the altitude from *A*. Let *E* and *F* be on a line through *D* such that *AE* is perpendicular to *BE*, *AF* is perpendicular to *CF*, and *E* and *F* are different from *D*. Let *M* and *N* be the midpoints of the line segments *BC* and *EF*, respectively. Prove that *AN* is perpendicular to *NM*.

Solution

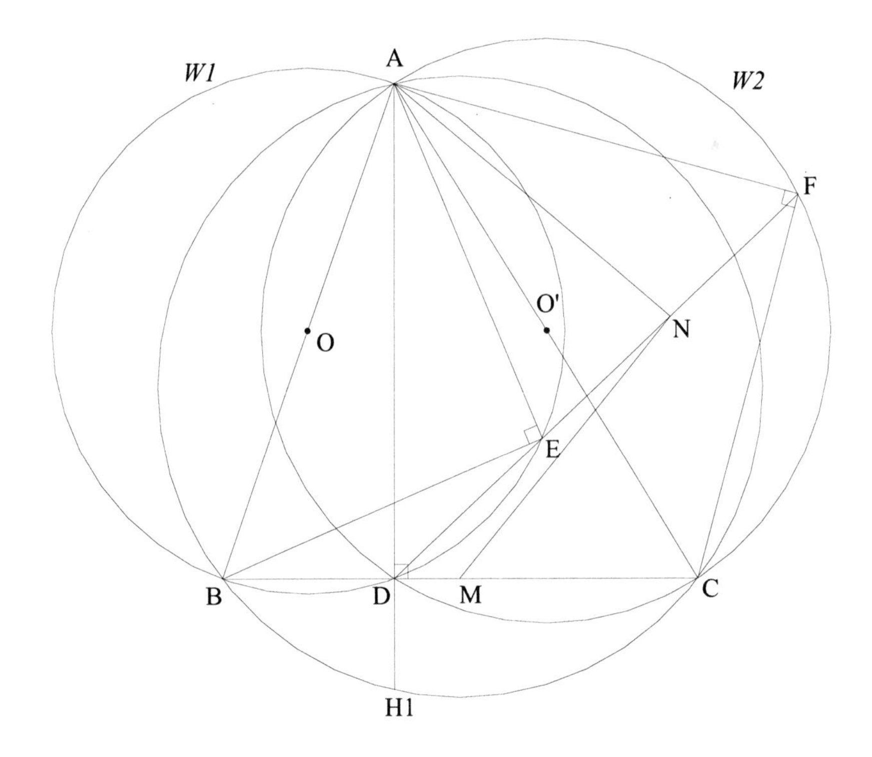

Let w1 and w2 be the circles with diameter AB and AC, respectively.

E is on w1 because AE is perpendicular to BE.
F is on w2 because AF is perpendicular to CF.

Let AN intercept w1 at Q and w2 at P. Point N is outside w1, and we have

$NE \times ND = NQ \times NA$ since N is midpoint of EF
$NF \times ND = NQ \times NA$ (i)

For w2, We have $NF \times ND = NP \times NA$ (ii)

From (i) and (ii), $NQ \times NA = NP \times NA$

or $NQ = NP$

Combining with MB = MC, we have MN || PC.

Since AP is perpendicular to PC and MN || PC therefore,
AN is perpendicular to NM.

Problem 4 of the Canadian Mathematical Olympiad 1969

Let ABC be an equilateral triangle, and P be an arbitrary point within the triangle. Perpendiculars PD, PE and PF are drawn to the three sides of the triangle. Show that, no mater where P is chosen.

$$(PD + PE + PF) / (AB + BC + CA) = 1/ 2\sqrt{3}$$

Solution

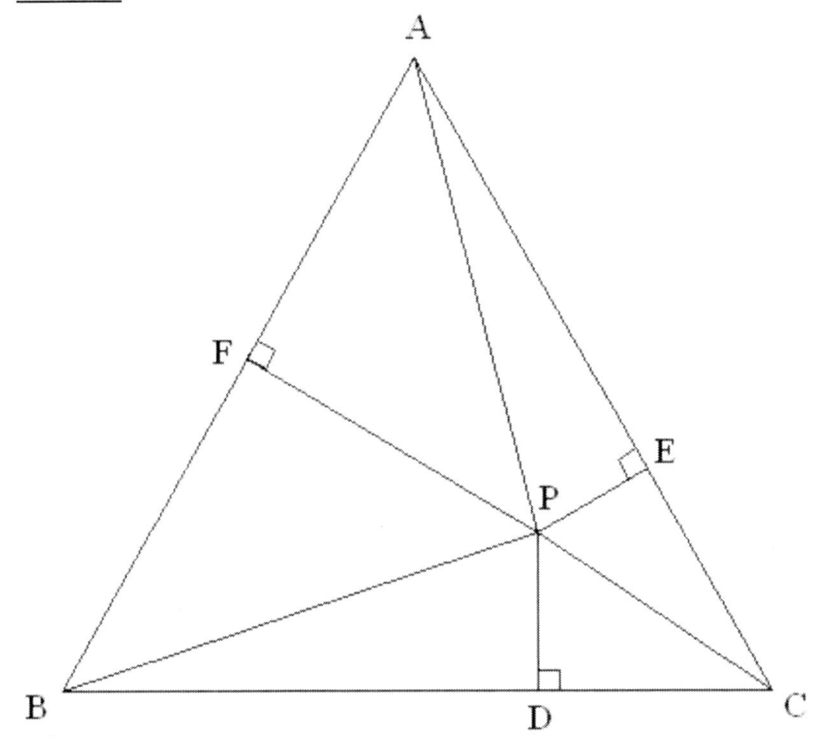

All altitudes of the equilateral triangle ABC are equal; let them be h and let a be the length of the side of ABC. Let (Ω) denote the area of shape Ω. We have $2(ABC) = a\,(PD + PE + PF) = ah$ or $(PD + PE + PF) = h$. Therefore,

$$(PD + PE + PF) / (AB + BC + CA) = h / (3a)$$

But $a^2 = h^2 + a^2/4$ or $h = a\sqrt{3}/2$ and $h/(3a) = 1/ 2\sqrt{3}$

Problem 4 of the Canadian Mathematical Olympiad 1970

a) Find all positive integers with initial digit 6 such that the integer formed by deleting this 6 is 1/25 of the original integer.
b) Show that there is no integer such that deletion of the first digit produces a result which is 1/35 of the original integer.

Solution

a) Let N = N0N1N2........Nn (n –> infinity) be such integers.
N0 = 6 and N = 6N1N2........Nn

By deleting the initial digit 6, we have M = N1N2........Nn and N / M = 25
and N – M = 60........0 (n number of 0's) = 24M or (N − M) /24 = M =

60........0 / 24 = 6 × 10........0 / (6 × 4) = 10........0 / 4 = 250...0 (n−2 number of 0's)

So N = 625, 6250, 62500, 625000, 6250000, 62500000, etc...

b) With the similar approach, assuming there are such integers N = N0N1N2........Nn
(n –> infinity) where

N0 = 1−9 (integers) and N = N0N1N2........Nn

By deleting the initial digit, we have M = N1N2........Nn and N / M = 35

and N – M = N00........0 (n number of 0's) = 34M or (N − M) /34 = M =

N00........0 / 34 = N0 × 10........0 / 34 and since N0 takes on the integer values of 1−9
N0 × 10........0 is not divisible by 34. Therefore, there are no such integers N as we assumed there were.

Problem 4 of Canadian Mathematical Olympiad 1971

Determine all real numbers a such that the two polynomials $x^2 + ax + 1$ and $x^2 + x + a$ have at least one root in common.

Solution

The roots for $x^2 + ax + 1 = 0$ are $x_{1\&2} = \dfrac{-a \pm \sqrt{a^2 - 4}}{2}$

The roots for $x^2 + x + a = 0$ are $x_{1\&2} = \dfrac{-1 \pm \sqrt{1 - 4a}}{2}$

Equating the roots of the two equations, we have

$-a \pm \sqrt{a^2 - 4} = -1 \pm \sqrt{1 - 4a}$ or $a^2 - 2a + 1 = a^2 - 4 + 1 - 4a \pm 2\sqrt{-4a^3 + a^2 + 16a - 4}$ or

$(2a + 4)^2 = 4(-4a^3 + a^2 + 16a - 4)$ or $a^3 - 3a + 2 = 0$ (i)

or $a^3 - 2a^2 + a + 2a^2 - 4a + 2 = (a + 2)(a^2 - 2a + 1) = (a + 2)(a - 1)^2 = 0$ or the solution for (i) are $a = -2$ and $a = 1$ and

When $a = -2$

the roots for $x^2 + ax + 1 = x^2 - 2x + 1 = (x - 1)^2 = 0$ is $x = 1$
the roots for $x^2 + x + a = x^2 + x - 2 = (x - 1)(x + 2) = 0$ are $x = 1$ and $x = -2$
so their common root is $x = 1$.

When $a = 1$

the roots for $x^2 + ax + 1 = x^2 + x + a = x^2 + x + 1$ so they are the same equation and

their roots are $x_{1\&2} = \dfrac{-1 \pm \sqrt{-3}}{2}$ (-3 is negative so they have no real roots).

Problem 4 of Canadian Mathematical Olympiad 1972

Describe a construction of a quadrilateral ABCD given:
(i) the lengths of all four sides;
(ii) that AB and CD are parallel;
(iii) that BC and DA do not intersect.

Solution

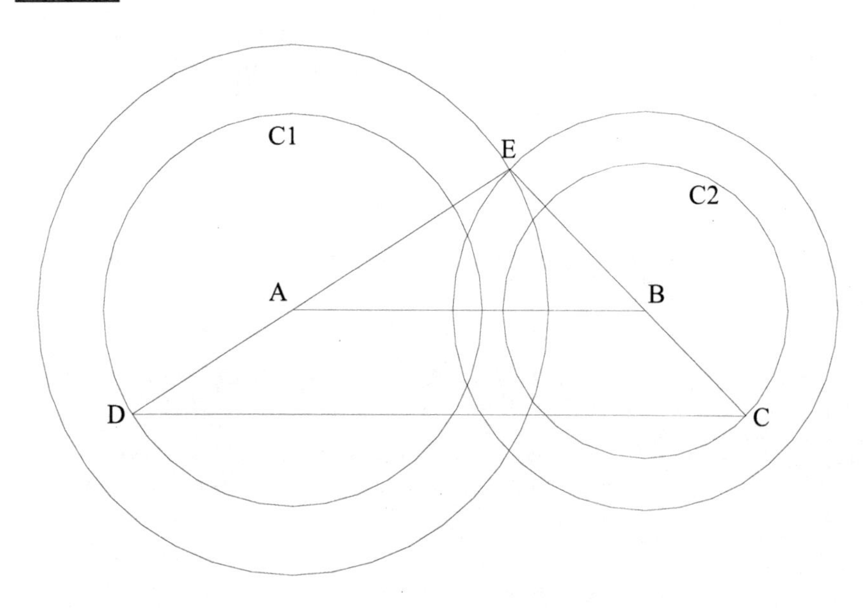

Lay side AB horizontally. Use A as a center, draw circle C1 with radius AD. Use B as a center and draw circle C2 with radius BC. Any horizontal line other than AB cutting through these two circles will have points D and C on them except that these arbitrary points C and D are not what we're looking for. To find C and D that form the given distance CD assume the extensions of DA and CB meet at E. We have

$AB/CD = EA/ED = EA/(EA + AD)$ or
$EA = AB \times AD/(CD - AB)$

Likewise, $AB/CD = EB/EC = EB/(EB + BC)$, or
$EB = AB \times BC/(CD - AB)$

Now use A and B as center and draw two circles with radii AE and BE, respectively. They will intercept at point E above segment AB.

Extend EA to meet circle C1 at D. The horizontal line passing through D will meet circle C2 at C where points E, B and C are collinear.

Problem 4 of the International Mathematical Olympiad 2007

In triangle ABC the bisector of angle BCA intersects the circumcircle again at R, the perpendicular bisector of BC at P, and the perpendicular bisector of AC at Q. The midpoint of BC is K and the midpoint of AC is L. Prove that the triangles RPK and RQL have the same area.

Solution:

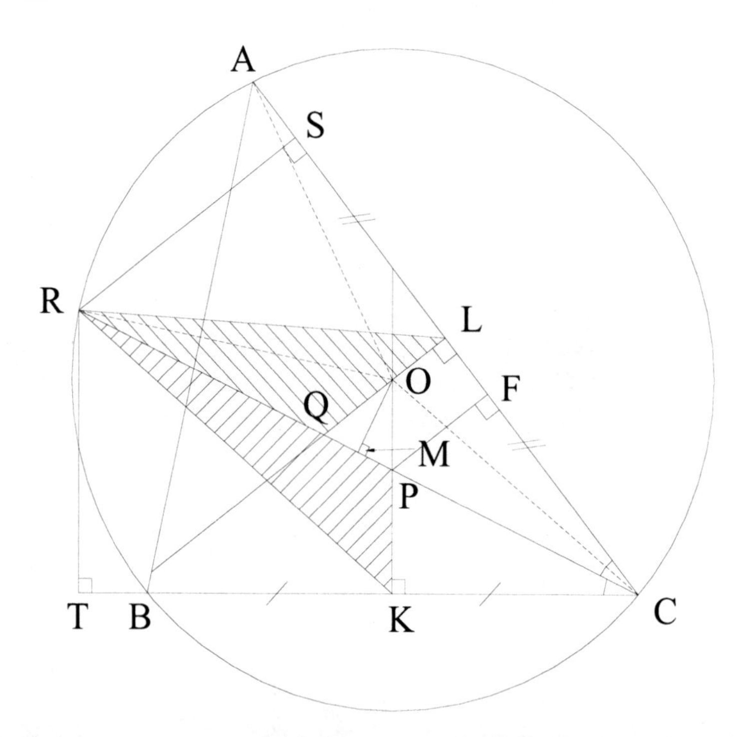

From R draw the two lines perpendicular to BC and AC and intersect them at T and S, respectively. From P also draw the perpendicular line and intersect AC at F.

To prove the two triangles RPK and RQL to have the same area, we need to prove: $TK \times PK = SL \times QL$ (i)

We know KPC and FPC are two identical triangles and the same for the two triangles TRC and SRC then TK = SF and PK = PF.

Equation (i) becomes
SF × PF = SL × QL which is what we need to prove or
PF / QL = SL / SF (ii)

but SL / SF = RQ / RP since all three lines RS, QL and PF are parallel then (ii) becomes

PF / QL = RQ / RP or QL / PF = RP / RQ (iii)
which we still need to
prove. Also note that QL / PF = QC / PC then (iii) becomes

QC / PC = RP / RQ (iv)
QC = QP + PC and RP = RQ + QP

Equation (iv) becomes 1 + QP / PC = 1 + QP / RQ that we need to prove, or QP / PC = QP / RQ
or PC = RQ (v) is what needs to be proven.

<u>Now let's prove it</u>
Note that O is the center of the circle. From O draw a line perpendicular to RC and intersect it at M.

$\angle MOP = \angle PCK$ (their lines are perpendicular) and
$\angle MOQ = \angle PCF$ for the same reason

since $\angle PCK = \angle PCF$ then angles MOP and MOQ are equal, and triangles MOP and MOQ are identical; therefore,
OQ = OP (v)

Note that OR = OC (ii) since O is the center of the circle and angles ROM and COM are equal then

$\angle ROQ = \angle COP$ (vi)

The three conditions (i), (ii) and (iii) make triangle ROQ and COP identical; therefore,

PC = RQ which is the condition (v) we need to prove.

Problem 4 of the International Mathematical Olympiad 2009

Let ABC be a triangle with AB = AC. The angle bisectors of $\angle$CAB and $\angle$ABC meet the sides BC and CA at D and E, respectively. Let K be the incenter of triangle ADC. Suppose that $\angle$BEK = 45°. Find all possible values of $\angle$CAB.

Solution

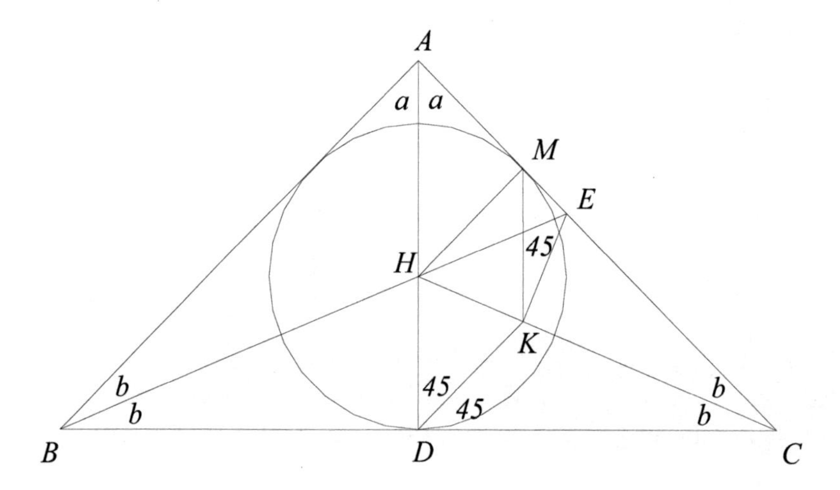

Extend CK to meet BE at H. H is the center of the incircle of triangle ABC. HM = HD

It's easily seen the two right triangles HDC and HMC are equal. Therefore, $\angle$HMK = $\angle$HDK = 45°.

There are two possibilities for $\angle$MHE. It's either 0° or positive.

Case 1: $\angle$MHE = 0°

Foot of B to AC is also the midpoint of AC, thus triangle ABC is equilateral and $\angle$CAB = 60°.

Case 2: $\angle$MHE > 0°

Problem also requires $\angle HEK = 45°$. Thus the circumcircle of triangle HMK will have point E on it.

Draw the circumcircle and we note that HE is the diameter since $\angle HME = 90°$. Point K is seen as the midpoint of the bottom arc HE. Therefore, $\angle KHE = 45°$.

$\angle BHC = 180° - \angle KHE = 135°$ or $\angle b = 22.5°$.

In triangle ABC we have

$\angle a + 2\angle b = 90°$.

Therefore, $\angle a = 45°$ or $\angle CAB = 90°$.

Problem 4 of USA Stanford Mathematical Tournament 2006

Simplify: $\dfrac{a^3}{(a - b)(a - c)} + \dfrac{b^3}{(b - a)(b - c)} + \dfrac{c^3}{(c - a)(c - b)}$

Solution

Expanding the expression, we have

$$\frac{a^3}{(a - b)(a - c)} + \frac{b^3}{(b - a)(b - c)} + \frac{c^3}{(c - a)(c - b)} =$$

$$\frac{a^3b - a^3c + b^3c - b^3a + c^3a - c^3b}{a^2b - a^2c + ac^2 - ab^2 + b^2c - bc^2}$$

Now perform a straight division, we have

$$\frac{a^3b - a^3c + b^3c - b^3a + c^3a - c^3b}{a^2b - a^2c + ac^2 - ab^2 + b^2c - bc^2} = a + b + c$$

Problem 4 of USA Mathematical Olympiad 1975

Two given circles intersect in two points P and Q. Show how to construct a segment AB passing through P and terminating on the two circles such that AP×PB is a maximum.

Solution

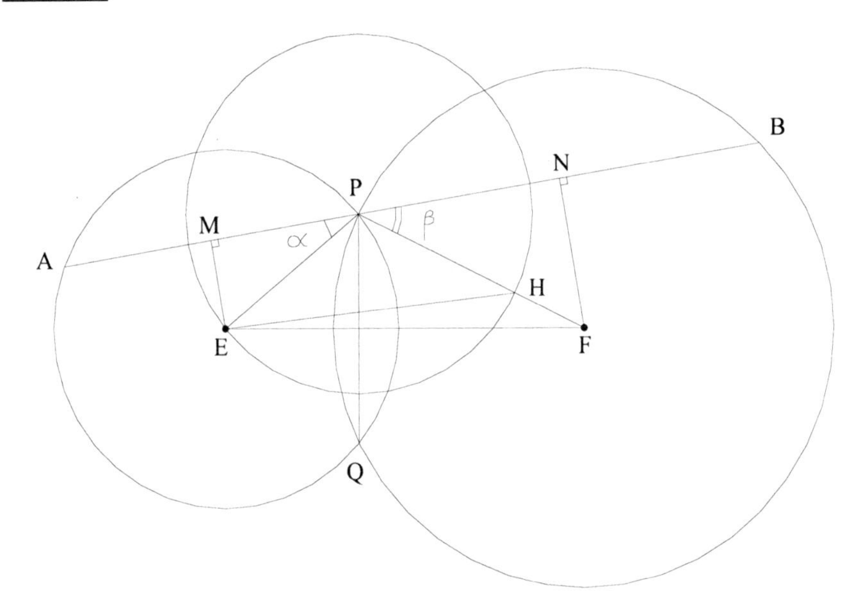

Let E and F be the centers of the small and big circles, respectively, and r and R be their respective radii.

Let M and N be the feet of E and F to AB, and $\alpha = \angle APE$ and $\beta = \angle BPF$

We have:

$$AP \times PB = 2r \cos \alpha \times 2r \cos \beta \ = 4 \text{ rR} \cos \alpha \times \cos \beta$$

AP × PB is maximum when the product $\cos \alpha \times \cos \beta$ is a maximum.

We have $\qquad \cos \alpha \times \cos \beta = ½ [\cos (\alpha + \beta) + \cos (\alpha - \beta)]$

But $\alpha + \beta = 180° - \angle EPF$ and is fixed, so is its $\cos(\alpha + \beta)$

So its maximum depends on $\cos(\alpha - \beta)$ which occurs when $\alpha = \beta$.
To draw the line AB:

Draw a circle with center P and radius PE to cut the radius PF at H. Draw the line parallel to EH passing through P. This line terminates the small and big circles at A and B, respectively.

<u>Further observation</u>

The problem below is derived from the above problem:

Two given circles intersect in two points P and Q. Show how to construct a segment AB passing through P and terminating on the two circles such that the ratio AP / PB equals the ratio of the two radii.

Problem 4 of the USA Mathematical Olympiad 1979

Show how to construct a chord FPE of a given angle A through a fixed point P within the angle A such that (1/FP) + (1/PE) is a maximum.

Solution

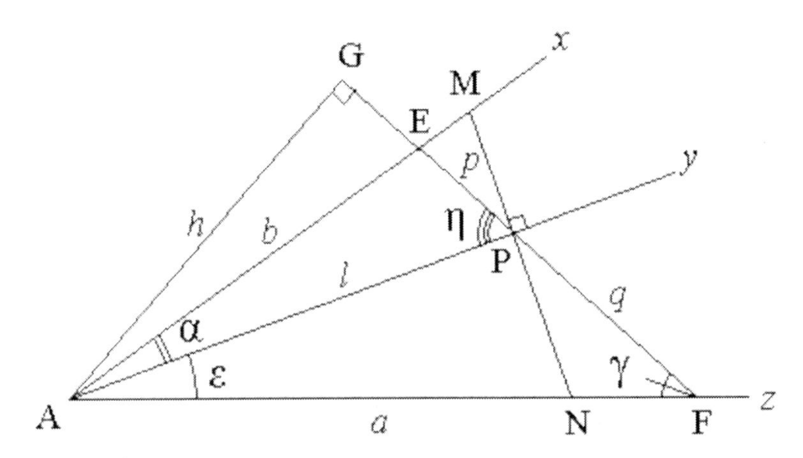

Let EP = p, FP = q, AF = a, AP = l, AE = b, ∠EAP = α, ∠FAP = ε, ∠EPA = η,
∠EFA = γ. Extend FE and from A draw a perpendicular line to intercept this extension
at G. Let AG = h.

$1/EP + 1/FP = 1/p + 1/q = (p + q)/(pq)$

Applying the law of the sine function, we obtain $(p + q)/\sin(\alpha + \varepsilon) = b/\sin\gamma$
and $b/p = \sin\eta/\sin\alpha$ or $(p + q)/(pq) = b\sin(\alpha + \varepsilon)/(pq\sin\gamma) = \sin(\alpha + \varepsilon)\sin\eta/(q\sin\gamma\sin\alpha)$

but α and ε are constants, so $1/p + 1/q$ is maximum when $\sin\eta/(q\sin\gamma)$ is maximum.

We also have $\sin\eta = h / l$ and $\sin\gamma = h / a$ and now $\sin\eta /(q \sin\gamma) = ha /(qlh) = a /(ql)$, but l is constant, so $a /(ql)$ is maximum is when ratio a/q is maximum

$a/q = \sin(180° - \eta) / \sin\varepsilon = \sin\eta / \sin\varepsilon$ and with angle ε fixed, a/q is maximum when $\sin\eta$ is maximum or equal to 1 when $\eta = 90°$ as line MN represents.

Problem 4 of the USA Mathematical Olympiad 2010

Let ABC be a triangle with $\angle A = 90°$. Points D and E lie on sides AC and AB, respectively, such that $\angle ABD = \angle DBC$ and $\angle ACE = \angle ECB$. Segments BD and CE meet at I. Determine whether or not it is possible for segments AB, AC, BI, ID, CI, IE to all have integer lengths.

Solution

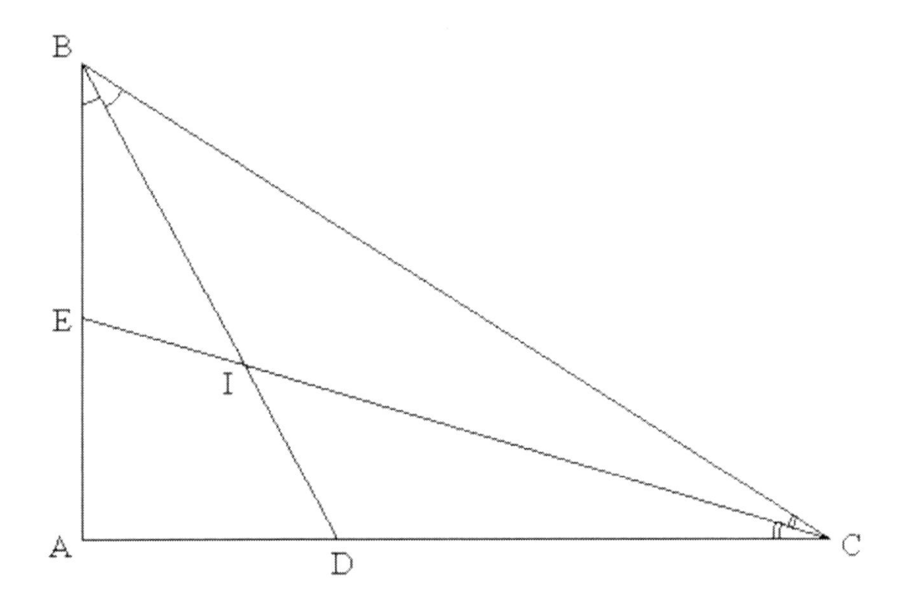

Apply the law of the cosine function, we have

$BC^2 = BI^2 + CI^2 - 2\, BI \times CI \cos\angle BIC$

But $\angle BIC = 180° - ½\,(180° - \angle A) = 135°$ and

$\cos\angle BIC = -\ ½\sqrt{2}$

The above equation becomes $BC^2 = BI^2 + CI^2 + \sqrt{2}\, BI \times CI$

Or $\sqrt{2}\, BI \times CI = BC^2 - BI^2 - CI^2$

Now assume that it is possible for segments AB, AC, BI, ID, CI and IE to all have integer lengths. BC^2 is then also an integer

because $BC^2 = AB^2 + AC^2$ which, in turn, requires $\sqrt{2}$ BI × CI to

be an integer. Since $\sqrt{2}$ is an irrational number, the product of $\sqrt{2}$ with an integer is not an integer. Therefore, our assumption was not possible, and it's not possible for segments AB, AC, BI, ID, CI and IE to all have integer lengths.

Problem 4 of the Vietnam Mathematical Olympiad 1962

Let be given a tetrahedron ABCD such that triangle BCD equilateral and AB = AC = AD. The height is h and the angle between two planes ABC and BCD is α. The point X is taken on AB such that the plane XCD is perpendicular to AB. Find the volume of the tetrahedron XBCD.

Solution

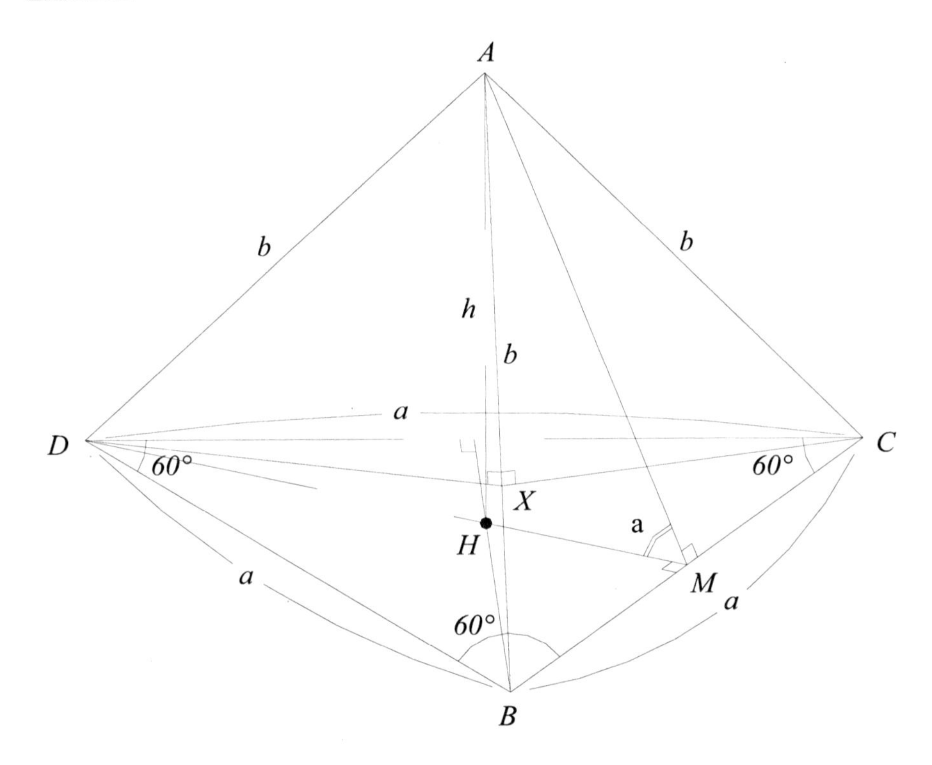

Let's find the area of triangle XDC denoted (XDC) and the length segment XB. Let a be the side length of triangle BDC and b = AB = AC = AD. Draw the altitude AM to BC and apply Pythagorean's theorem, we have

$AM^2 = b^2 - \dfrac{a^2}{4}$ or $AM = \dfrac{1}{2}\sqrt{4b^2 - a^2}$ and $BC^2 = CX^2 + BX^2$ or $a^2 = CX^2 + BX^2$.

Since M is the midpoint of BC and ABC is an isosceles triangle with AB = AC and triangle BCD is equilateral, $\alpha = \angle AMD$.

We also have $\tan\angle ABC = \frac{AM}{BM} = \frac{CX}{BX} = \frac{\sqrt{4b^2 - a^2}}{a}$

Now solving the two equations, we have $BX = \frac{a^2}{2b}$,

$CX = \frac{a}{2b}\sqrt{4b^2 - a^2}$

Now apply Heron's formula for (XDC), taking into account that CX = DX, we have

$(XDC) = \sqrt{s(s\text{-}a)(s\text{-}CX)^2}$ where $s = \frac{a}{2} + CX = \frac{a}{2} + \frac{a}{2b}\sqrt{4b^2 - a^2}$,

$s - CX = \frac{a}{2}$, $s - a = -\frac{a}{2} + \frac{a}{2b}\sqrt{4b^2 - a^2}$

After some computations, $(XDC) = \frac{a^2}{4b}\sqrt{3b^2 - a^2}$

The volume of the tetrahedron XBCD, by definition, is

$V = \frac{1}{3}(XDC) \times BX = \frac{a4}{24b^2}\sqrt{3b^2 - a^2}$

Furthermore, $h^2 = AM^2 - HM^2 = AM^2 - (DM/3)^2 = \frac{3b^2 - a^2}{3}$ or

$h = \sqrt{\frac{3b^2 - a^2}{3}}$. The volume is now $V = \frac{a4h\sqrt{3}}{24b^2}$ (i)

But $\tan\alpha = \frac{h}{HM} = \frac{6h}{a\sqrt{3}}$, or $a = \frac{6h}{\tan\alpha\sqrt{3}}$

And $b^2 = AM^2 + BM^2 = h^2 + HM^2 + \frac{a^2}{4} = h^2 + \frac{a^2}{12} + \frac{a^2}{4} = h^2 + \frac{a^2}{3}$

Or $b^2 = h^2 + \frac{4h^2}{\tan^2\alpha}$

Substituting the values of a and b^2 into (i), we have

The volume of the tetrahedron XBCD $= \frac{6h^3\sqrt{3}}{\tan^2\alpha\,(\tan^2\alpha + 4)}$

Problem 4 of the Vietnam Mathematical Olympiad 1986

Let ABCD be a square of side 2a. An equilateral triangle AMB is constructed in the plane through AB perpendicular to the plane of the square. A point S moves on AB such that SB = x. Let P be the projection of M on SC and E, O be the midpoints of AB and CM, respectively.

(a) Find the locus of P as S moves on AB.
(b) Find the maximum and minimum lengths of SO.

Solution

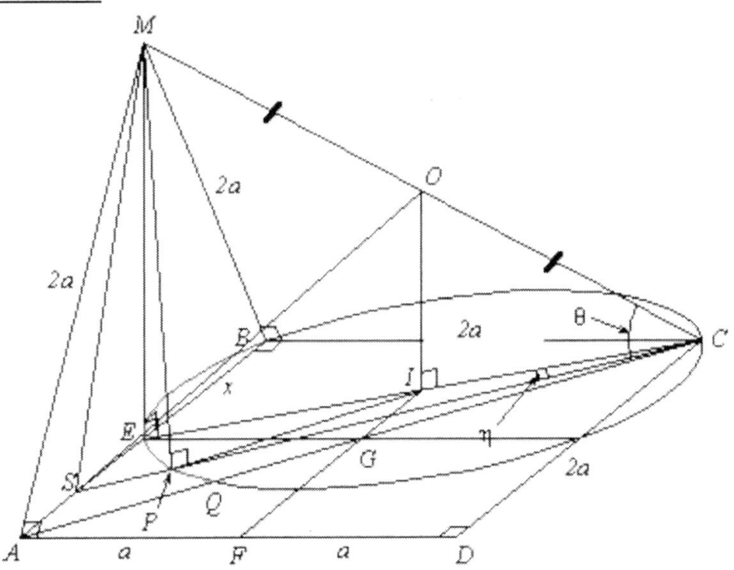

(Since x is already taken by the problem, × is being used to denote multiplication). Let $\theta = \angle MCP$, $\eta = \angle ECP$ and I be the midpoint of CE. The lengths of the other segments are calculated to be CA $= CM = 2\sqrt{2}a$, $CE = a\sqrt{5}$, $IC = \frac{1}{2}CE = \frac{1}{2}a\sqrt{5}$,

$ME = a\sqrt{3}$, $SE = x - a$, $MS = \sqrt{(x - a)^2 + 3a^2}$, $SC = \sqrt{x^2 + 4a^2}$

a) Apply the law of the cosine function, we have

$MS^2 = CM^2 + SC^2 - 2CM \times SC \cos\theta$, or

$(x - a)^2 + 3a^2 = 8a^2 + x^2 + 4a^2 - 4\sqrt{2}a \times \sqrt{x^2 + 4a^2} \cos\theta$, or

$$\cos\theta = \frac{x + 4a}{2\sqrt{2(x^2 + 4a^2)}}$$

but $\cos\theta = \frac{CP}{CM}$; therefore, $CP = 2\sqrt{2}a \frac{x + 4a}{2\sqrt{2(x^2 + 4a^2)}} = \frac{a(x + 4a)}{\sqrt{x^2 + 4a^2}}$

Again, the law of the cosine function gives us
$SE^2 = CE^2 + SC^2 - 2\ CE \times SC \cos\eta$, or

$(x - a)^2 = 5a^2 + x^2 + 4a^2 - 2a\sqrt{5(x^2 + 4a^2)} \cos\eta$, or

$$\cos\eta = \frac{x + 4a}{\sqrt{5(x^2 + 4a^2)}}$$

$IP^2 = IC^2 + CP^2 - 2\ IC \times CP \cos\eta$, or

$$IP^2 = \frac{5a^2}{4} + \frac{a^2(x + 4a)^2}{x^2 + 4a^2} - a\sqrt{5}\frac{a(x + 4a)}{\sqrt{x^2 + 4a^2}} \times \frac{x + 4a}{\sqrt{5(x^2 + 4a^2)}} = \frac{5a^2}{4}, \text{ or}$$

$$IP = \frac{1}{2}a\sqrt{5}$$

which is a constant, and the locus of P is part of the circle that has its center at I and radius of $\frac{1}{2}a\sqrt{5}$ that passes through point E and is from B to Q where Q is the intersection of the circle and CA.

b) Since I and O are the midpoints of CE and CM, respectively, IO || ME, and the plane containing the three points M, C and E is perpendicular with the plane of the square, IO is then perpendicular with CE and $SO^2 = IO^2 + SI^2$.

But $IO = \frac{1}{2}ME = \frac{1}{2}a\sqrt{5}$ is fixed; the extreme values of SO depend on SI. As S moves on AB, SI is minimum when S is at the midpoint of EB (SI = a) and is maximum when S is at A when $SI^2 = AI^2 = AF^2 + FI^2$ where F is the midpoint of AD. Let G be the midpoint of AC.

$SI^2 = AF^2 + (FG + GI)^2 = a^2 + (a + \frac{a}{2})^2 = \frac{13a^2}{4}$, and

$SO^2_{max} = \frac{5a^2}{4} + \frac{13a^2}{4} = \frac{9a^2}{2}$, or $SO_{max} = \frac{3a}{\sqrt{2}}$, and

$SO^2_{min} = \frac{5a^2}{4} + a^2 = \frac{9a^2}{4}$, or $SO_{min} = \frac{3a}{2}$

Problem 5 of the International Mathematical Olympiad 2004

In a convex quadrilateral ABCD the diagonal BD does not bisect the angles ABC and CDA. The point P lies inside ABCD and satisfies $\angle PBC = \angle DBA$ and $\angle PDC = \angle BDA$. Prove that ABCD is a cyclic quadrilateral if and only if AP = CP.

Solution

1. Assume point B is already on the circle

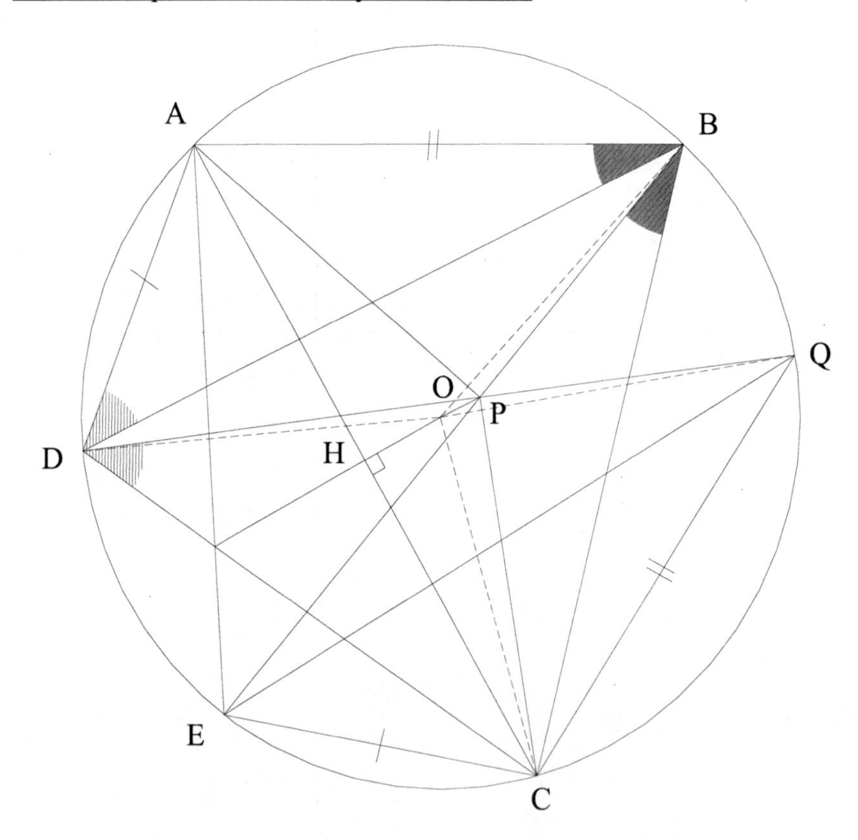

Extend DP to intercept the circle at Q and BP to intercept the circle at E.

Consider two quadrilaterals ABPD and CQPE:

Since $\angle ABD = \angle EBC$ -> AD = EC

$\angle ABE = \angle DBC = \angle DQC$
$\angle DPB = \angle EPQ$
$\angle ADQ = \angle BDC = \angle BEC$

Therefore, $\angle DAB = \angle ECQ$ since the sum of the angles of a quadrilateral is 360°.

Two triangles DAB and ECQ are identical since $\angle DAB = \angle ECQ$,

AD = EC and
AB = CQ implies DB = EQ

Therefore, triangles DPB and EPQ are also identical (two angles on each side of DB and EQ are equal which gives us PB = PQ.

Therefore, triangles ABP and CQP are identical since AB = CQ, PB = PQ and the two angles $\angle ABP$ and $\angle CQP$ are equal which implies

$$AP = PC.$$

<u>2. Assume AP = PC and prove ABCD is a cyclic quadrilateral</u>

<u>Proof using contradiction</u>

Assume we already have triangle ADC with A, B and C on the circle. From C draw a line that
intercepts the circle at E and that CE = AD.

Point P can be chosen anywhere and draw a line to connect E and P and extend it to cut the circle at B to satisfy the first condition $\angle ABD = \angle PBC$ since CE = AD.

Now let point O be the center of the circle. We note that OP is the center line of symmetry of AD and CE. Extend DP to intercept the circle at Q.

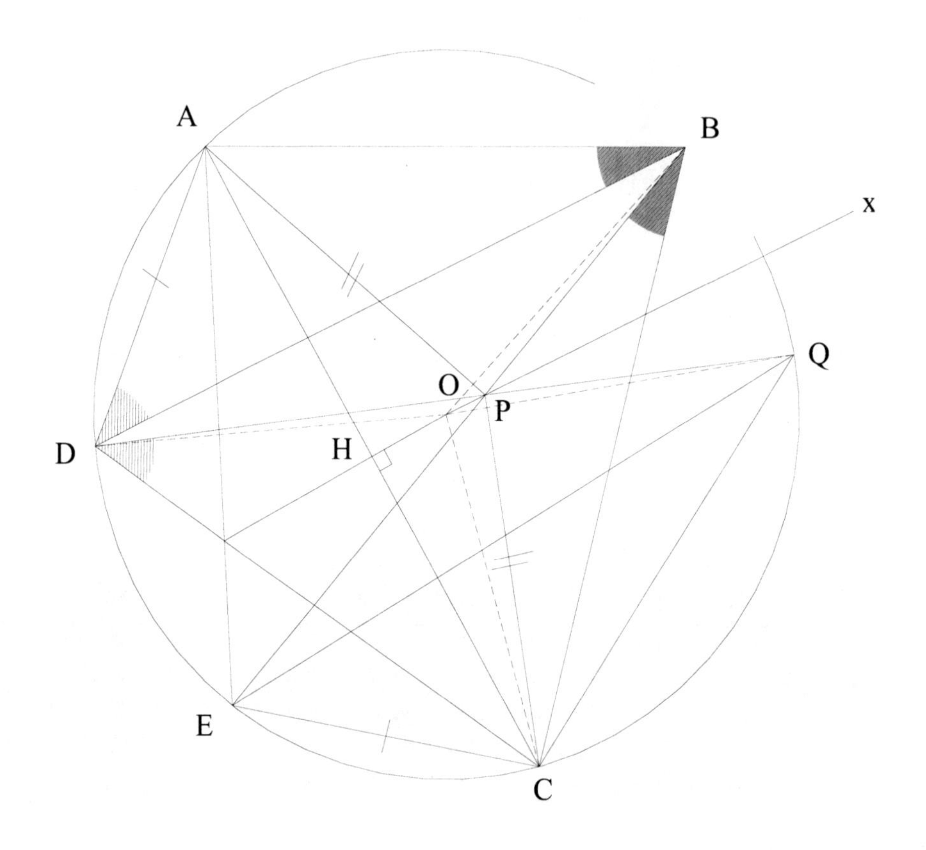

Assume AP ≠ PC (contradict with fact) then CQ ≠ AB since Ox (extension of OP) is the line of symmetry. Since Q is on the circle and AB ≠ CQ therefore, $\angle ADB \neq \angle PDC$ implies B is not symmetrical of Q via Ox. Therefore, B is not on the circle.

So P has to be on Ox for B to be on the circle or AP = PC then ABCD is a cyclic quadrilateral.

Problem 5 of USA Mathematical Olympiad 1990

An acute-angled triangle ABC is given in the plane. The circle with diameter AB intersects altitude CC' and its extension at points M and N , and the circle with diameter AC intersects altitude BB' and its extensions at P and Q. Prove that the points M;N; P;Q lie on a common circle.

Solution

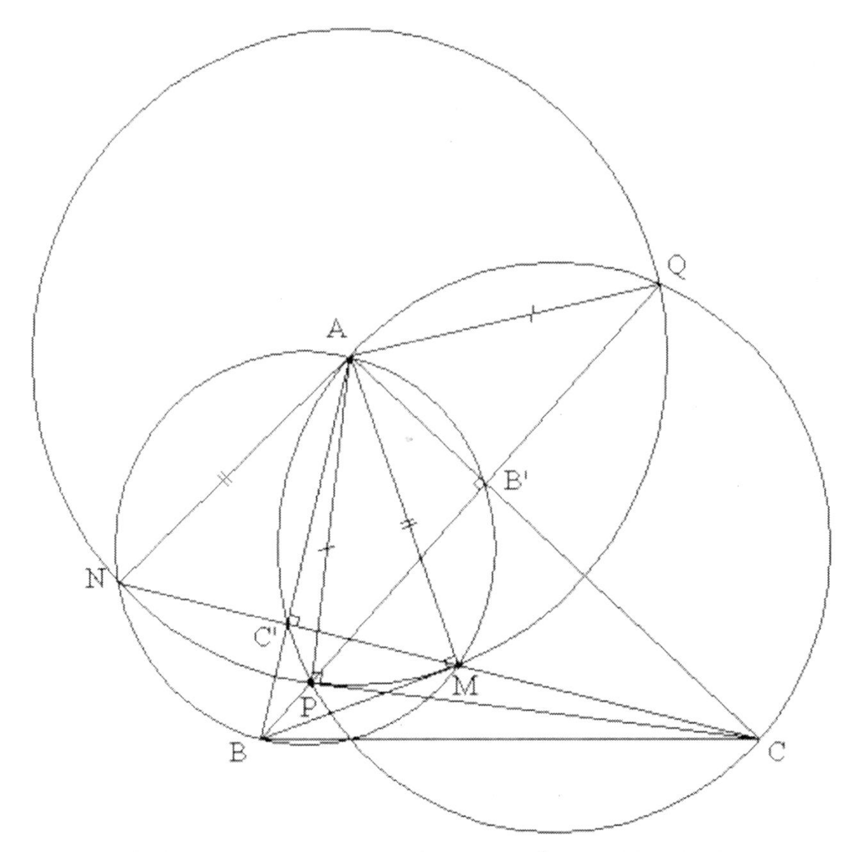

We already have AP = AQ and AN = AM
To prove the 4 points M, N, P and Q to lie on a common circle, we only need to prove AM = AP

Because AC is the diameter, angle APC = 90°, we have $AP^2 = AC^2 - PC^2$.

or $\quad AP^2 = AC'^2 + CC'^2 - PC^2 \qquad$ (i)

$\quad AM^2 = AC'^2 + C'M^2 \qquad$ (ii)

Substituting AC^2 from (i) to (ii), we have

$$AM^2 = AP^2 - CC'^2 + PC^2 + C'M^2$$

So to prove AM = AP, we need to prove $CC'^2 = C'M^2 + PC^2$ (iii)

We have $PC^2 = B'C^2 + PB'^2 = B'C^2 + PB' \times B'Q = B'C^2 + AB' \times B'C = B'C\,(B'C + AB') = B'C \times AC = CM \times CN$ (iv)

Substituting PC^2 from (iv) to (iii), we then need to prove

$CC'^2 = C'M^2 + CM \times CN \qquad$ (v)

But $CC' = CM + MC'$, and (v) becomes

$CM^2 + 2\,CM \times MC' + C'M^2 = C'M^2 + CM \times CN \qquad$ (vi)

Or we now need to prove

$CM^2 + 2\,CM \times MC' = CM \times CN \qquad$ (vii)

or $\quad CM\,(CM + 2MC') = CM \times CN$

Because $C'M = C'N$ therefore, $CM + 2MC' = CN$ and the problem is solved.

The points M, N, P and Q lie on a common circle with center A and radius AP.

Problem 5 of USA Mathematical Olympiad 1996

Triangle ABC has the following property: there is an interior point P such that $\angle PAB = 10°$, $\angle PBA = 20°$, $\angle PCA = 30°$, and $\angle PAC = 40°$. Prove that triangle ABC is isosceles.

Solution

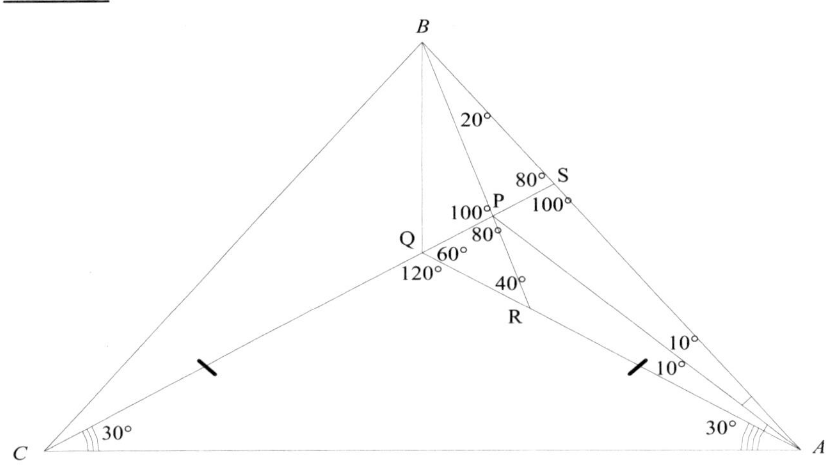

Extend CP to meet AB at S. From A draw a line to meet the extension of BP at R and CP at Q such that $\angle QAP = 10°$. We have BR = AR and

$\angle BRQ = 40°$, $\angle QAC = 30°$, $\angle AQC = 120°$, $\angle PQR = 60°$, $\angle QPR = 80°$, $\angle PSA = 100°$, $\angle QPB = 100°$ and $\angle BSP = 80°$

We need to prove the two triangles QSA and QPB are similar since if they are similar we have $\angle QBP = \angle QAS = 20°$ and $\angle QBA = 40°$, $\angle BQA = 120° = \angle BQC$ and the two triangles BQC and BQA are congruent and thus BC = BA and the triangle ABC is isosceles.
To prove those two triangles that have the $\angle QSA = \angle QPB = 100°$ similar, we need to prove

$$QP / QS = PB / SA \qquad \text{(i)}$$

Since AP is bisector of $\angle QAS$, we have QP / QA = PS / SA (ii)
QP / QA = PS / SA = QS / (QA + SA) or
QP / QS = QA / (QA + SA)

Combining with (i), we now need to prove QA / (QA + SA) = PB / SA = (QA – PB) / QA = (QR + PR) / QA (since BR = AR), or we need to prove

$$(QR + PR) / QA = PB / SA \qquad \text{(iii)}$$

Using the law of the since function, we have

$QP / \sin 40° = QR / \sin 80° = PR / \sin 60° = (QR + PR)/(\sin 60° + \sin 80°)$ and $PS / \sin 20° = PB / \sin 80°$

or $QP = (QR + PR) \sin 40° /(\sin 60° + \sin 80°)$ and
$PS = PB \sin 20° / \sin 80°$

Substituting QP and PS to (ii), it becomes

$[(QR + PR) / QA] \times [\sin 40° / (\sin 60° + \sin 80°)] = [PB / SA] \times [\sin 20° / \sin 80°]$

so now we have to prove
$$\sin 40° / (\sin 60° + \sin 80°) = \sin 20° / \sin 80° \qquad \text{(iv)}$$

or $\sin 20° / \sin 80° = (\sin 40° - \sin 20°) / \sin 60°$,

or $\sin 10° / \sin 30° = \sin 20° / \sin 80°$,

or $\sin 10° \sin 80° = \sin 30° \sin 20°$,

or $\frac{1}{2} (\cos 70° - \cos 90°) = \cos 60° \cos 70°$,

or $\frac{1}{2} = \cos 60°$ which is obvious!

Problem 6 of Austrian Mathematical Olympiad 2008

We are given a square ABCD. Let P be different from the vertices of the square and from its center M. For a point P for which the line PD intersects the line AC, let E be this intersection. For a point P for which the line PC intersects the line DB, let F be this intersection. All those points P for which E and F exist are called acceptable points.

Determine the set of acceptable points for which the line EF is parallel to AD.

Solution

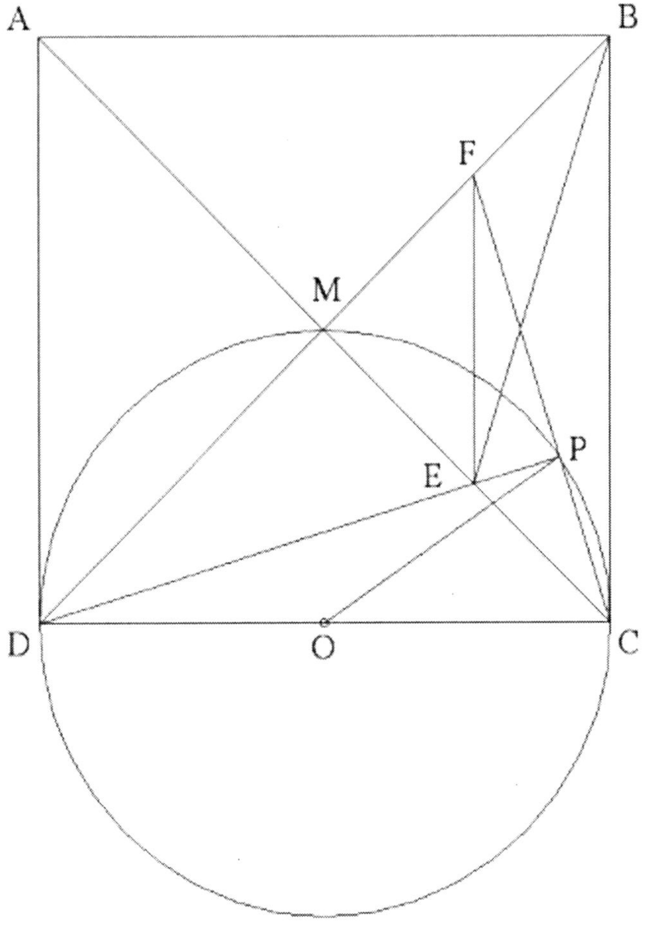

Link EB. Since ABCD is a square and AC is the perpendicular bisector of BD, and E is on AC, ED = EB and $\angle EDB = \angle EBD$.

Furthermore, since EF || BC, EFBC is an isosceles trapezoid and $\angle ECF = \angle EBD$, or $\angle ECF = \angle EDB$

We also have $\angle FCB + \angle ECF = 45°$, or

$$\angle FCB + \angle EDB = 45°, \text{ or}$$
$$\angle FCB = \angle EDC, \text{ and}$$
$$FC \perp DP, \text{ or } \angle DPC = 90°.$$

So all the acceptable points form a circle with center O being the midpoint of DC and the diameter equals to the side length of the square ABCD.

Problem 6 of the Australian Mathematical Olympiad 2010

Prove that $\sqrt[3]{6+\sqrt[3]{845}+\sqrt[3]{325}}+\sqrt[3]{6+\sqrt[3]{847}+\sqrt[3]{539}} =$
$\sqrt[3]{4+\sqrt[3]{245}+\sqrt[3]{175}}+\sqrt[3]{8+\sqrt[3]{1859}+\sqrt[3]{1573}}$

Solution

Observe that $845 = 13^2 \times 5$, $325 = 5^2 \times 13$
$847 = 11^2 \times 7$, $539 = 7^2 \times 11$
$245 = 7^2 \times 5$, $175 = 5^2 \times 7$
$1859 = 13^2 \times 11$, $1573 = 11^2 \times 13$

and $6 + \sqrt[3]{845}+\sqrt[3]{325} = (\sqrt[3]{\frac{13}{3}})^3 + 3\,(\sqrt[3]{\frac{13}{3}})^2$

$\sqrt[3]{\frac{5}{3}} + 3\sqrt[3]{\frac{13}{3}}\,(\sqrt[3]{\frac{5}{3}})^2 + (\sqrt[3]{\frac{5}{3}})^3 = [\sqrt[3]{\frac{13}{3}}+\sqrt[3]{\frac{5}{3}}\,]^3$

or $\sqrt[3]{6+\sqrt[3]{845}+\sqrt[3]{325}} = \sqrt[3]{\frac{13}{3}}+\sqrt[3]{\frac{5}{3}}$

Similarly, $\sqrt[3]{6+\sqrt[3]{847}+\sqrt[3]{539}} = \sqrt[3]{\frac{11}{3}}+\sqrt[3]{\frac{7}{3}}$

$\sqrt[3]{4+\sqrt[3]{245}+\sqrt[3]{175}} = \sqrt[3]{\frac{7}{3}}+\sqrt[3]{\frac{5}{3}}$

$\sqrt[3]{8+\sqrt[3]{1859}+\sqrt[3]{1573}} = \sqrt[3]{\frac{13}{3}}+\sqrt[3]{\frac{11}{3}}$

Therefore, $\sqrt[3]{6+\sqrt[3]{845}+\sqrt[3]{325}}+\sqrt[3]{6+\sqrt[3]{847}+\sqrt[3]{539}} =$
$\sqrt[3]{4+\sqrt[3]{245}+\sqrt[3]{175}}+\sqrt[3]{8+\sqrt[3]{1859}+\sqrt[3]{1573}}$

Problem 6 of Belarusian Mathematical Olympiad 2000

The equilateral triangles ABF and CAG are constructed in the exterior of a right-angled triangle ABC with $\angle C = 90°$. Let M be the midpoint of BC. Given that MF = 11 and MG = 7, find the length of BC.

Solution

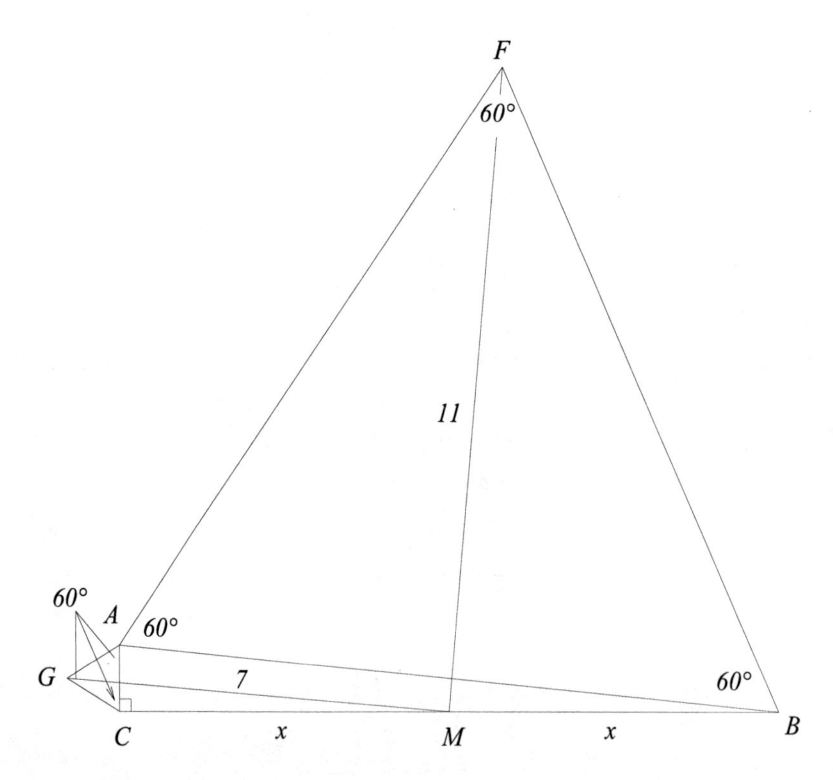

Let $x = \frac{BC}{2}$, AB = b and AC = c.

Apply the law of cosines, we have

$GM^2 = x^2 + c^2 - 2xc\cos(\angle ACB + 60°)$, or

$GM^2 = x^2 + c^2 - 2xc\cos 150°$

and $\quad FM^2 = x^2 + b^2 - 2xb\cos(\angle ABC + 60°)$

Expanding those two equations with MG = 7 and MF = 11,
$\cos(\angle ABC + 60°) = \cos\angle ABC\cos 60° - \sin\angle ABC\sin 60°$,

$\cos 60° = \frac{1}{2}$, $\sin 60° = \frac{\sqrt{3}}{2}$, $\quad \cos 150° = -\frac{\sqrt{3}}{2}$

and observe the Pythagorean's theorem, we have

$49 = c^2 + x^2 + xc\sqrt{3}$

$121 = b^2 - x^2 + xc\sqrt{3}$

$b^2 = c^2 + 4x^2$

Solving for x, we obtain x = 6 or BC = 12.

Problem 6 of the Belarusian Mathematical Olympiad 2004

Circles S1 and S2 meet at points A and B. A line through A is parallel to the line through the centers of S1 and S2 and meets S1 again at C and S2 again at D. The circle S3 with diameter CD meets S1 and S2 again at P and Q, respectively. Prove that lines CP, DQ, and AB are concurrent.

Solution

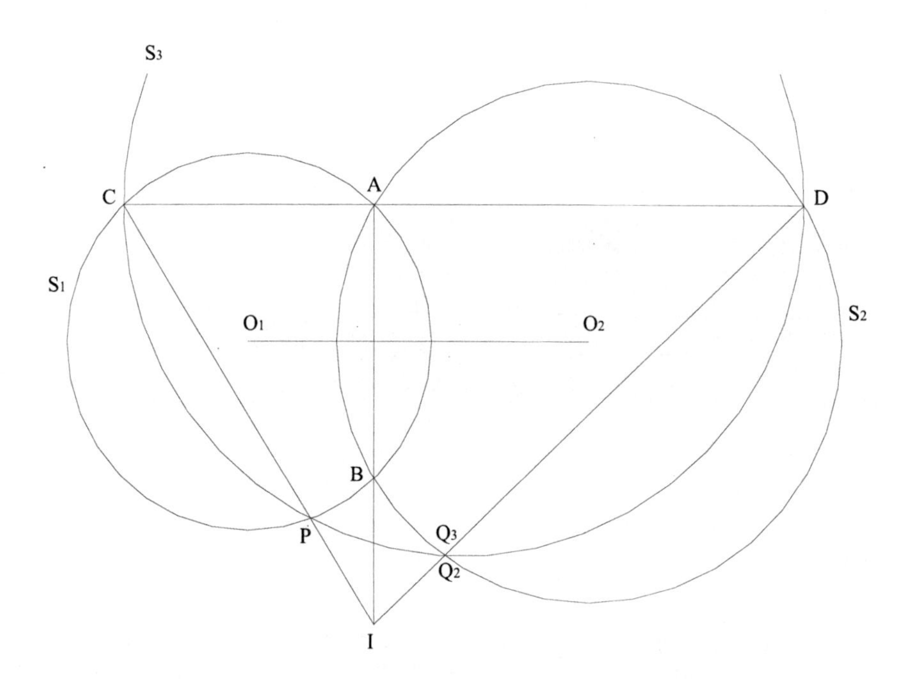

Extend CP to meet the extension of AB at I. Link ID to meet S3 at Q3 and S2 at Q2.

We have

IP × IC = IQ3 × ID since (C, P, Q3 and D are on S3) (i)

IP × IC = IB × IA since (C, P, B and A are on S1) (ii)

IQ2 × ID = IB × IA since (A, B, Q2 and D are on S2) (iii)

From (ii) and (iii) $IP \times IC = IQ2 \times ID$ (iv)

From (i) and (iv) $IQ3 \times ID = IQ2 \times ID$

or $IQ3 = IQ2$

But S3 and S2 only meet at a single point Q, therefore, $Q3 \equiv Q2 \equiv Q$, or the three CP, DQ, and AB are concurrent.

Note: Line CD does not have to parallel to the line through the centers of S1 and S2. The result is still the same regardless.

Problem 6 of the Canadian Mathematical Olympiad 1970

Given three non-collinear points A, B, C, construct a circle with center C such that the tangents from A and B to the circle are parallel.

Solution

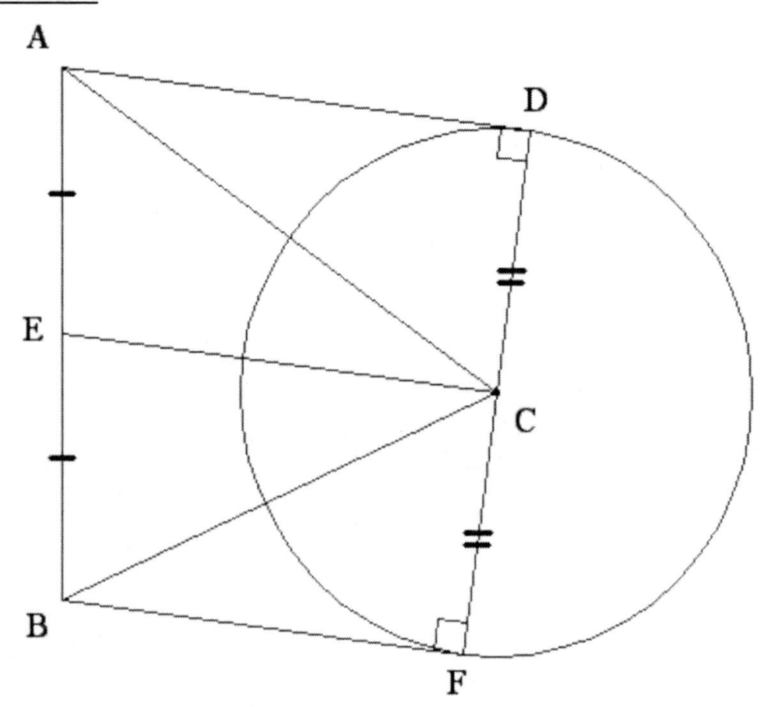

Let D and F be the points where the parallel lines from A and B tangent the circle, respectively, and E be the midpoint of AB.
We have CD = CF.

Since $CD \perp AD$, $CF \perp BF$ and $AD \parallel BF$, the three points D, C and F are collinear, and
ABFD is a trapezoid with $CE = \frac{1}{2}(AD + BF)$.

To draw the circle, link EC. From A and B draw two parallel lines and draw the circle with the center C to tangent these parallel lines at D and F, respectively.

Problem 7 of the Canadian Mathematical Olympiad 1971

Let n be a five digit number (whose first digit is non-zero) and let m be the four digit number formed from n by deleting its middle digit. Determine all n such that n / m is an integer.

Solution

Let $n = abcde$ where a, b, c, d and e are positive integers from 0 to 9 and $a \neq 0$.

We then have $m = abde$ and

$$n = 10000a + 1000b + 100c + 10d + e$$
$$m = 1000a + 100b + 10d + e$$

If $n / m = k$ is an integer, we have

$$10000a + 1000b + 100c + 10d + e = 1000ak + 100bk + 10dk + ek \quad \text{(i)}$$

Now assume $k > 10$ or $k = 10 + p$ where p is a positive integer; (i) becomes

$$10000a + 1000b + 100c + 10d + e = 10000a + 1000b + 100d + 10e + 1000ap + 100bp + 10dp + ep \quad \text{(ii)}$$

Now let's find the possible value for p. We have

$$p = \frac{100c - 90d - 9e}{1000a + 100b + 10d + e}$$

but since $a \neq 0$ and b, c, d and e are all non-negative integers, the denominator is then ≥ 1000 and the numerator is less than 1000, so $p < 1$, and $k > 10$ is not possible.

Similarly, if $k < 10$, $p = (90d + 9e - 100c)/(1000a + 100b + 10d + e)$

With the same argument $k < 10$ is not a possibility. Therefore, $k = 10$.

Substituting $k = 10$ into (i), we have $100c = 90d + 9e$ which requires product 9e to be a multiple of 10 which is not possible. This equation has the only solution

$c = d = e = 0$. So n = ab000 where a and b are positive integers where a = 1–>9 and
b = 0–>9. So numbers n are

10000, 11000, 12000, 13000, 14000, 15000, 16000, 17000, 18000, 19000,
20000, 21000, 22000, 23000, 24000, 25000, 26000, 27000, 28000, 29000,
30000, 31000, 32000, 33000, 34000, 35000, 36000, 37000, 38000, 39000,
40000, 41000, 42000, 43000, 44000, 45000, 46000, 47000, 48000, 49000,
50000, 51000, 52000, 53000, 54000, 55000, 56000, 57000, 58000, 59000,
60000, 61000, 62000, 63000, 64000, 65000, 66000, 67000, 68000, 69000,
70000, 71000, 72000, 73000, 74000, 75000, 76000, 77000, 78000, 79000,
80000, 81000, 82000, 83000, 84000, 85000, 86000, 87000, 88000, 89000,
90000, 91000, 92000, 93000, 94000, 95000, 96000, 97000, 98000, 99000

a total of 90 numbers.

Problem 7 of Irish Mathematical Olympiad 1994

Let p, q, r be distinct real numbers which satisfy the equations

$q = p(4 - p)$ (i)
$r = q(4 - q)$ (ii)
$p = r(4 - r)$ (iii)

Find all possible values of $p + q + r$.

Solution

Adding the three equations, we have
$3(p + q + r) = p^2 + q^2 + r^2$ (iv)

Without loss of generality, assume $p > q > r$ and $p = q + m$;
$q = r + n$

We then have $p = r + m + n$
where m and n are real numbers, and $m \neq 0$, $n \neq 0$, and $m \neq n$.

We then have $p + q + r = 3r + m + 2n$

Substituting p and q into (i), we have
$3(3r + m + 2n) = (r + m + n)^2 + (r + n)^2 + r^2$ or

$3r^2 + (2m + 4n - 9)r + m^2 + 2n^2 + 2mn - 3m - 6n = 0$

$r_{1\&2} = \frac{1}{6}[9 - 2m - 4n \pm\sqrt{-8m^2 - 8n^2 - 8mn + 81}\ ] = \frac{1}{6}[9 - 2m - 4n$
$\pm\sqrt{-8m^2 - 8n^2 - 8mn + 81}\ ]$

And $p + q + r = 3r + m + 2n = \frac{9}{2} \pm \frac{1}{2}$
$\sqrt{9^2 - 4[(m + n)^2 + m^2 + n^2]}$

Observe that $(m + n)^2 + m^2 + n^2 \in (0, \frac{9^2}{2^2}]$; therefore,

$p + q + r \in (0, 9)$

Problem 8 of the Canadian Mathematical Olympiad 1970

Consider all line segments of length 4 with one end-point on the line y = x and the other end-point on the line y = 2x. Find the equation of the locus of the midpoints of these line segments.

Solution

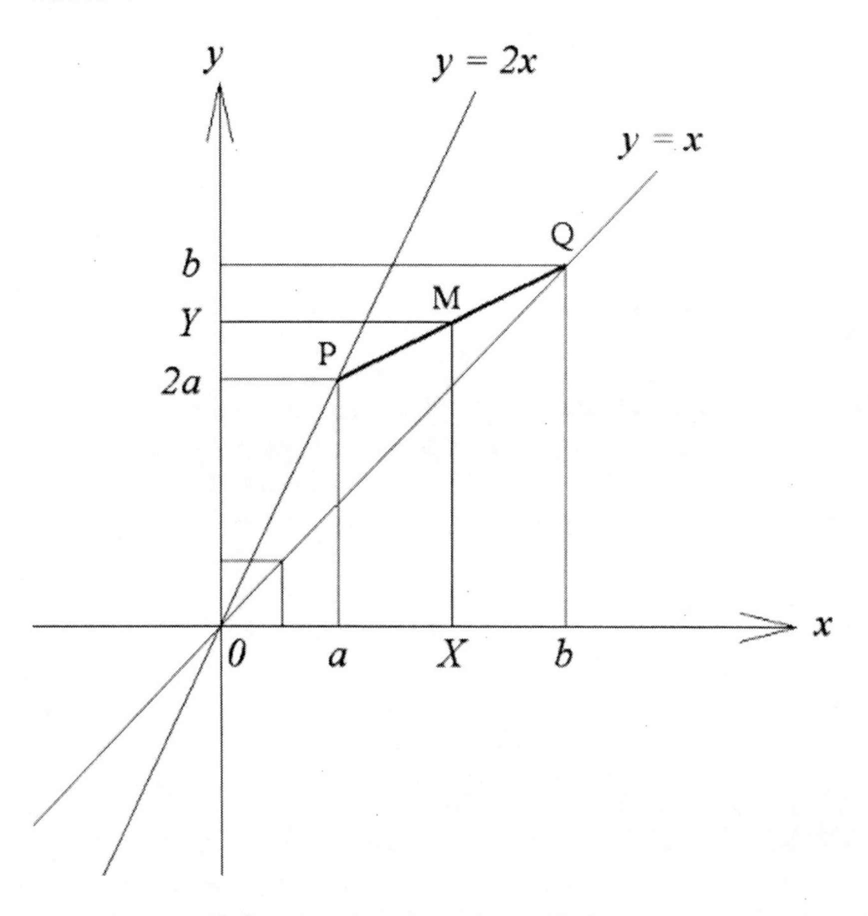

Let the segment be PQ, P on y = 2x and Q on y = x, its midpoint be M. We have PQ = 4 and M(X, Y) meaning X, and Y are coordinates of point M on the x and y axes, respectively, Q(b, b) since point Q is on y = x line, and P(a, 2a) because it's on y=2x line. To use the least numbers of unknowns possible, let's pick the half segment MQ for the calculation. We have

$MQ^2 = (b - X)^2 + (b - Y)^2 = (PQ/2)^2 = 4$ (i)

Besides, $X = (a + b)/2$ or $a/2 = X - b/2$

and $Y = (2a + b)/2 = X + a/2 = X + X - b/2 = 2X - b/2$ or $b/2 = 2X - Y$ or $b = 4X - 2Y$

Substituting b into (i), we have

$$13Y^2 - 36XY + 25X^2 - 4 = 0$$

or $Y = 18X/13 \pm \sqrt{52 - X^2}/13$

when $X \leq 0$ $Y = 18X/13 + \sqrt{52 - X^2}/13$

when $X \geq 0$ $Y = 18X/13 - \sqrt{52 - X^2}/13$

Note that the locus only goes from point N $(\sqrt{52},\ 18\sqrt{52}/13)$ to

N' $(-\sqrt{52},\ -18\sqrt{52}/13)$.

Problem 8 of the Canadian Mathematical Olympiad 1971

A regular pentagon is inscribed in a circle of radius r. P is any point inside the pentagon. Perpendiculars are dropped from P to the sides, or the sides produced, of the pentagon.

a) Prove that the sum of the lengths of these perpendiculars is constant.
b) Express this constant in terms of the radius r.

Solution

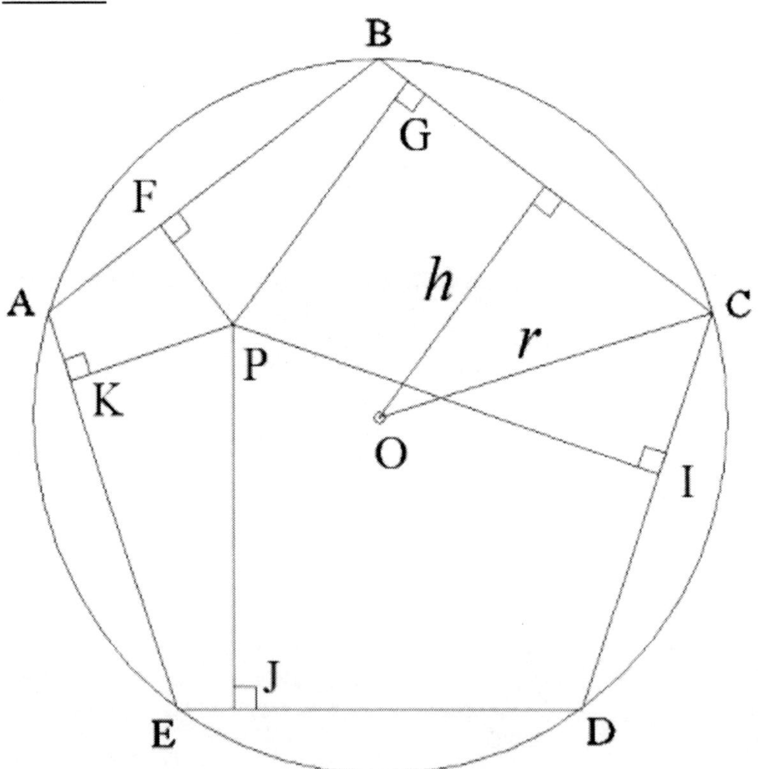

Let ABCDE be such regular pentagon (5 sides equaled) and a be its side length (a = AB = BC = CD = DE = EA). Now let F, G, I, J, K be the feet from P to the sides AB, BC, CD, DE and EA, respectively.

a) The area of the pentagon is

$$\tfrac{1}{2}\,(FP\times AB + GP\times BC + IP\times CD + JP\times ED + KP\times EA) =$$
$$\tfrac{1}{2}\,a\,(FP + GP + IP + JP + KP) = \tfrac{1}{2}\,ah \times 5$$

or that the sum of the lengths of these perpendiculars is a constant

$$c = FP + GP + IP + JP + KP = 5h$$

b) Consider triangle BOC, $\angle BOC = 360°/ 5 = 72°$, or
$\angle BCO = 90° - \tfrac{1}{2}\angle BOC = 54°$.

but $\sin 54° = h/r$, therefore, the constant $c = 5h = 5r \sin 54° = 4.05r$

Problem 9 of the Auckland Mathematical Olympiad 2009

Through the incenter I of triangle ABC a straight line is drawn intersecting AB and BC at points M and N, respectively, in such a way that the triangle BMN is acute-angled. On the side AC the points K and L are chosen such that $\angle ILA = \angle IMB$ and $\angle IKC = \angle INB$. Prove that AC = AM + KL + CN.

Solution

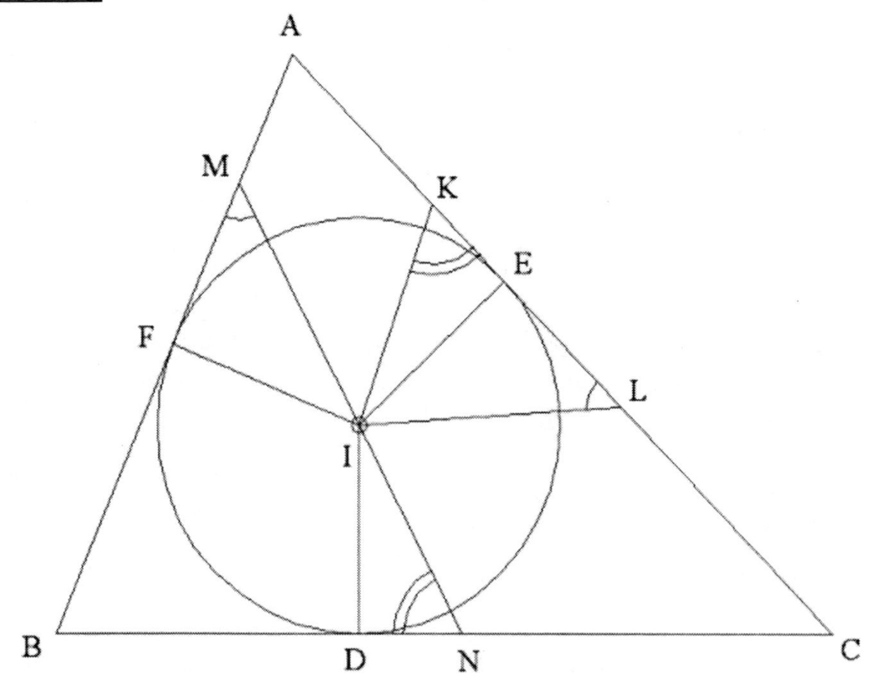

Let the incircle tangent with BC, AC and AB at D, E and F, respectively.

Triangles IDN and IEK are congruent since all their angles are equal and ID = IE = radius of the incircle. Therefore, KE = ND. Likewise, triangles IFM = triangle IEL and LE = MF.

AC = AF + CD = AM + MF + CN + ND = AM + LE + CN + KE = AM + KL + CN.

Problem 9 of Canadian Mathematical Olympiad 1970

Let f(n) be the sum of the first n terms of the sequence

0, 1, 1, 2, 2, 3, 3, 4, 4, 5, 5, 6, 6, ...

a) Give a formula for f(n).
b) Prove that $f(s + t) - f(s - t) = st$ where s and t are positive integers and $s > t$.

Solution

a) Since $f(i) = 0$ does not belong to the sequence, $f(n + 1)$ is really f(n) in a sense, so we have to subtract 1 from the n sequence, and we have

$$f(n) = \frac{n-1}{2}\left(\frac{n-1}{2} + 1\right)$$

b) Therefore, $f(s + t) = \frac{s+t-1}{2} \times \frac{s+t-1}{2}$, and

$$f(s - t) = \frac{s-t-1}{2} \times \frac{s-t+1}{2} = st$$

Problem 9 of Irish Mathematical Olympiad 1994

Let w, a, b, c be distinct real numbers with the property that there exist real numbers x, y, z for which the following equations hold:

$x + y + z = 1$ (i)
$xa^2 + yb^2 + zc^2 = w^2$ (ii)
$xa^3 + yb^3 + zc^3 = w^3$ (iii)
$xa^4 + yb^4 + zc^4 = w^4$ (iv)

Express w in terms of a, b, c.

Solution

Multiplying both sides of (i) by a^2, a^3 and a^4, we have
$xa^2 + ya^2 + za^2 = a^2$ (v)
$xa^3 + ya^3 + za^3 = a^3$ (vi)
$xa^4 + ya^4 + za^4 = a^4$ (vii)

Subtracting (ii) from (v), (iii) from (vi) and (iv) from (vii), we have
$y(a^2 - b^2) + z(a^2 - c^2) = a^2 - w^2$ (viii)
$y(a^3 - b^3) + z(a^3 - c^3) = a^3 - w^3$ (ix)
$y(a^4 - b^4) + z(a^4 - c^4) = a^4 - w^4$ (x)

Now multiplying both sides of (viii) by $a^2 + b^2$, we have
$y(a^4 - b^4) + z(a^2 - c^2)(a^2 + b^2) = (a^2 - w^2)(a^2 + b^2)$ (xi)

Subtracting (x) from (xi), we have
$z(a^2 - c^2)(b^2 - c^2) = (a^2 - w^2)(b^2 - w^2)$ (xii)

Multiplying both sides of (viii) by $\dfrac{a^2 + ab + b^2}{a + b}$, we have

$$y(a^3 - b^3) + z(a^2 - c^2)\frac{a^2 + ab + b^2}{a + b} = (a^2 - w^2)\frac{a^2 + ab + b^2}{a + b}$$
(xiii)

Subtracting (xiii) from (ix), we have

$$z(a-c)\left[(a+c)\frac{a^2+ab+b^2}{a+b}-a^2-ac-c^2\right]=$$

$$(a-w)\ \left[(a+w)\frac{a^2+ab+b^2}{a+b}-a^2-aw-w^2\right] \qquad \text{(xiv)}$$

Now dividing (xiv) by (xii), we have

$$\left[(a+c)\frac{a^2+ab+b^2}{a+b}-a^2-ac-c^2\right]/\left[(a+c)(b^2-c^2)\right]=$$

$$\left[(a+w)\frac{a^2+ab+b^2}{a+b}-a^2-aw-w^2\right]/\left[(a+w)(b^2-w^2)\right] \qquad \text{(xv)}$$

Expanding (xv) and canceling the same terms, we have

$$\frac{ab^2+b^2c-ac^2-bc^2}{(a+c)(b^2-c^2)}=\frac{ab^2+b^2w-aw^2-bw^2}{(a+w)(b^2-w^2)} \qquad \text{or}$$

$$(ab+ac+bc)w^2-c^2(a+b)w-abc^2=0$$

Solving for w, we have

$$w_{1\&2}=\frac{c}{2(ab+ac+bc)}\left[c(a+b)\pm\sqrt{c^2(a+b)^2+4ab(ab+ac+bc)}\right]$$

But $c^2(a+b)^2+4ab(ab+ac+bc)=(ac+bc+2ab)^2$; therefore,

$$w_{1\&2}=\frac{c}{2(ab+ac+bc)}\left[c(a+b)\pm(ac+bc+2ab)\right]$$

$$w=c,\ \text{ or }\ w=-\frac{abc}{ab+ac+bc}\ \text{ which requires } ab+ac+bc\neq 0$$

Substitute $w = c$ into (ii) and reverse the above processes; multiply both sides of (ii) by c and subtract it from (iii), etc... We found that when $w = c$, $a = b = c$ which is not allowed by the problem.

Therefore, the only possible solution possible $w=-\dfrac{abc}{ab+ac+bc}$.

Problem 9 of the Middle European Mathematical Olympiad 2009

Let ABCD be a parallelogram with $\angle BAD = 60°$ and denote by E the intersection of its diagonals. The circumcircle of triangle ACD meets the line BA at K≠A, the line BD at P≠D and the line BC at L≠C. The line EP intersects the circumcircle of triangle CEL at points E and M. Prove that triangles KLM and CAP are congruent.

Solution

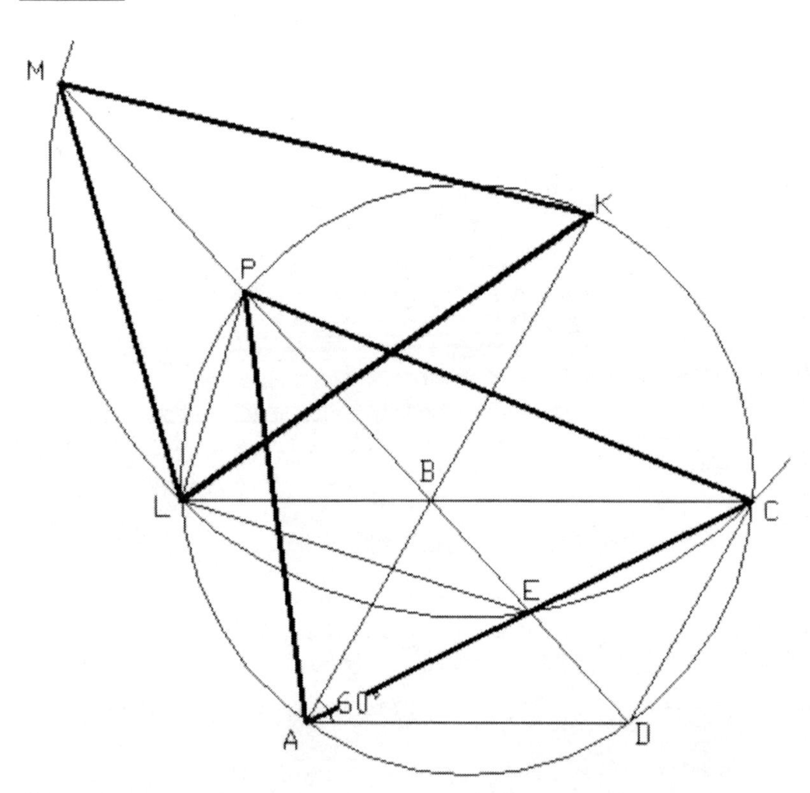

Since K, L, A, C and M, L, E, C are concyclic, we have $\frac{KL}{AC} = \frac{BL}{AB}$, $\frac{AP}{CD} = \frac{AE}{DE}$ and $\frac{ML}{EC} = \frac{BL}{BE}$ or $\frac{ML}{AP} = \frac{EC \text{ x } BL \times DE}{BE \times AE \times CD}$

but since E is the intersection of the diagonals of the parallelogram ABCD, BE = DE, AE = EC.

It follows that $\frac{ML}{AP} = \frac{BL}{CD} = \frac{BL}{AB} = \frac{KL}{AC}$ (i)

We also have $\angle LPD = \angle LCD = \angle BAD = 60°$

Now chase the angle
$\angle KLM = \angle KLP + \angle MLP = \angle KLP + 60° - \angle LMP = \angle KLP + 60° - \angle LME = \angle KLP + 60° - \angle LCE = \angle KLP + \angle ACD =$ $\angle KLP + \angle KLC$ (since AB || CD and KC = AD) $= \angle CLP =$ $\angle CAP$ (subtending arc CP).

Combining with (i), the two triangles KLM and CAP are similar.

Furthermore, since DL = AC (diagonals of isosceles trapezoid ADCL) = DK (diagonals of isosceles trapezoid ADCK).

But since $\angle BAD = 60°$subtends arc DK, it follows that KDL is an equilateral triangle, and KL = DK = AC.

This makes the two already similar triangles KLM and CAP congruent.

Problem 2 of the Auckland Mathematical Olympiad 2009

Is it possible to write the number $1^2 + 2^2 + 3^2 + \ldots + 12^2$ as a sum of 11 distinct squares?

Solution

We note that $5^2 + 12^2 = 25 + 144 = 169 = 13^2$ and

$1^2 + 2^2 + 3^2 + \ldots + 12^2 = 1^2 + 2^2 + 3^2 + 4^2 + 6^2 + 7^2 + 8^2 + 9^2 + 10^2 + 11^2 + 13^2$

which is the sum of 11 distinct squares.

Also note that $3^2 + 4^2 = 5^2$, the expression can be written as a sum of 10 distinct squares

$1^2 + 2^2 + 3^2 + \ldots + 12^2 = 1^2 + 2^2 + 5^2 + 6^2 + 7^2 + 8^2 + 9^2 + 10^2 + 11^2 + 13^2$.

<u>Problem 2 of the Australian Mathematical Olympiad 2008</u>

Let ABC be an acute-angled triangle, and let D be the point on AB (extended if necessary) such that AB and CD are perpendicular. Further, let tA and tB be the tangents to the circumcircle of ABC through A and B, respectively, and let E and F be the points on tA and tB, respectively, such that CE is perpendicular to tA and CF is perpendicular to tB.
Prove that $CD/CE = CF/CD$.

<u>Solution</u>

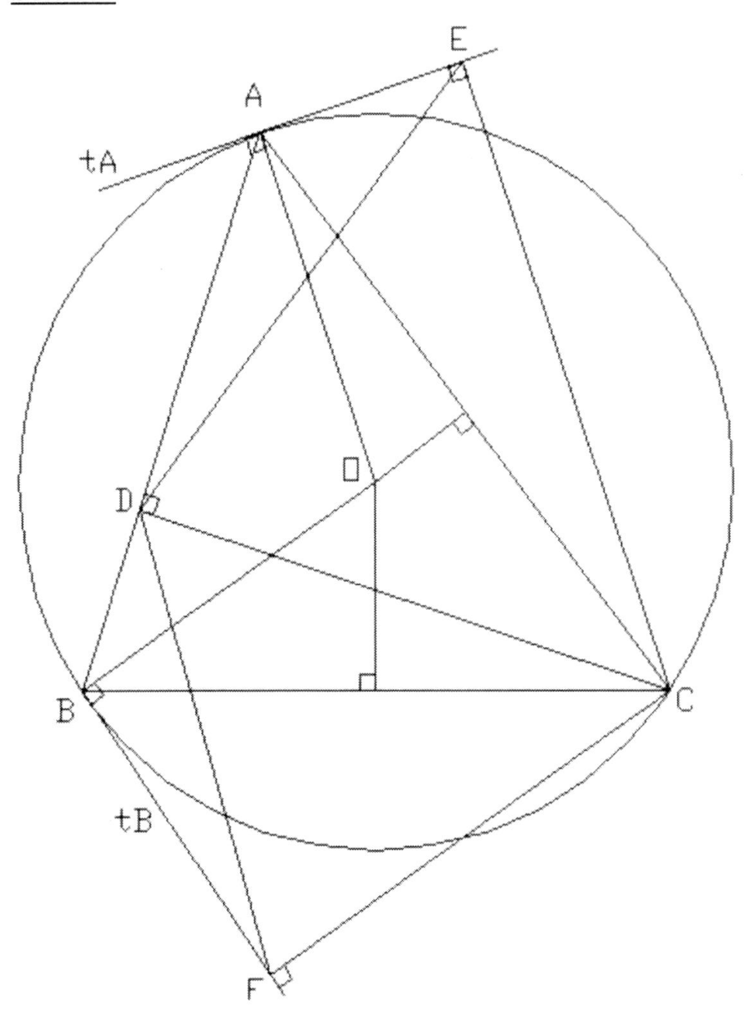

We see that ADCE and BDCF are both cyclic which cause
$\angle DFC = \angle ABC$ (i)
and $\angle DEC = \angle BAC$ (ii)
and
$\angle BAE + \angle DCE = \angle ABF + \angle DCF = 180°$ but $\angle BAE =$ $\angle ABF$ (both subtend larger arc AB); therefore, $\angle DCE = \angle DCF$.

However, $\angle ACB$ subtends smaller arc AB; hence $\angle ACB =$ $\angle DCE = \angle DCF$ since $\angle ACB$ also combines with $\angle BAE$ to be 180°.

Now with the addition of (i), the triangles ABC and DFC are similar because their respective angles are equal.

Similarly, combined with (ii), the two triangles ABC and EDC are also similar for the same reason.

Therefore, triangles DFC and EDC are similar, and

$CD/CE = CF/CD.$

Problem 2 of Belarusian Mathematical Olympiad 1997 Category D

Points D and E are taken on side CB of triangle ABC, with D between C and E, such that $\angle BAE = \angle CAD$. If AC < AB, prove that AC · AE < AB · AD.

Solution

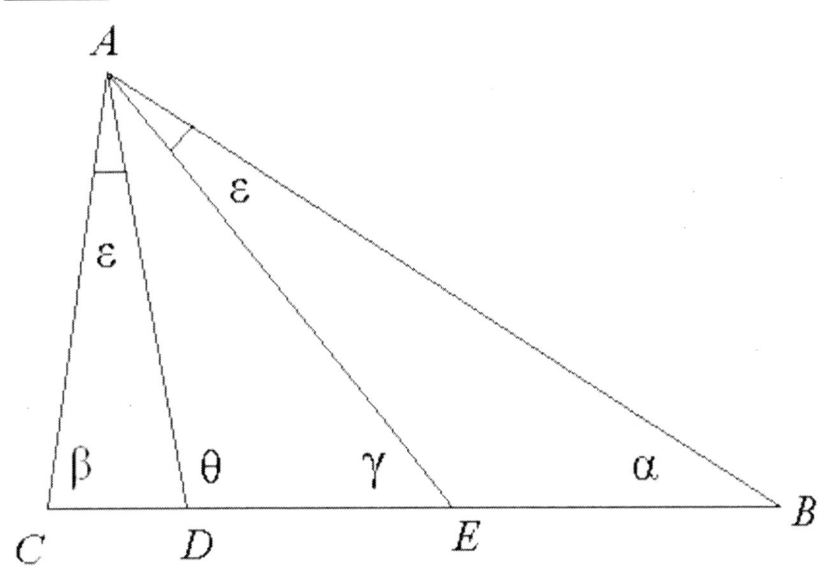

Let $\alpha = \angle ABC$, $\beta = \angle ACB$, $\gamma = \angle AEC$, $\theta = \angle ADB$ and $\varepsilon = \angle BAE = \angle CAD$.

To prove AC · AE < AB · AD, we need to prove

$$\frac{AC}{AB} < \frac{AD}{AE} \qquad \text{(i)}$$

Apply the law of sine function, (i) becomes $\dfrac{\sin\alpha}{\sin\beta} < \dfrac{\sin\gamma}{\sin\theta}$

or $\quad \sin\alpha \sin\theta < \sin\beta \sin\gamma \quad$ or

$-\frac{1}{2}[\cos(\alpha + \theta) - \cos(\alpha - \theta)] < -\frac{1}{2}[\cos(\beta + \gamma) - \cos(\beta - \gamma)]$, or

$$\cos(\alpha + \theta) - \cos(\alpha - \theta) > \cos(\beta + \gamma) - \cos(\beta - \gamma) \qquad \text{(ii)}$$

but $\alpha + \theta = 180° - \angle BAD = 180° - \angle DAE - \varepsilon = \beta + \gamma$ and (ii) becomes $-\cos(\alpha - \theta) > -\cos(\beta - \gamma)$,

or

$$\cos(\alpha - \theta) < \cos(\beta - \gamma), \text{ or } \quad \cos(\theta - \alpha) < \cos(\gamma - \beta) \qquad \text{(iii)}$$

But we are given AC < AB which causes $\beta > \alpha$ and $\theta = \beta + \varepsilon > \alpha + \varepsilon = \gamma$, or $\theta - \alpha > \varepsilon$ and $\gamma - \beta < \varepsilon$ or $\theta - \alpha > \gamma - \beta$ and (iii) is a reality.

Problem 2 of the Canadian Mathematical Olympiad 1969

Determine which of the two numbers $\sqrt{c+1} - \sqrt{c}$, $\sqrt{c} - \sqrt{c-1}$ is greater for any $c \geq 1$.

Solution

Since $c \geq 1$, we always have $c > \sqrt{c^2 - 1}$,

or $\quad 2c > 2\sqrt{c^2 - 1}$,

or $\quad 4c > 2c + 2\sqrt{c^2 - 1} = [\sqrt{c+1} + \sqrt{c-1}\,]^2$,

or $\quad 2\sqrt{c} > \sqrt{c+1} + \sqrt{c-1}$,

or $\quad \sqrt{c} - \sqrt{c-1} > \sqrt{c+1} - \sqrt{c}$

Problem 2 of the Canadian Mathematical Olympiad 1970

Given a triangle ABC with angle A obtuse and with altitudes of length h and k as shown in the diagram, prove that $a + h \geq b + k$. Find under what conditions $a + h = b + k$.

Solution

The two triangles AEC and BDC are similar, we have $a / k = b / h = a - b / k - h$

In figure 1, we have $b > h$ or $b / h > 1$ or $a - b / k - h > 1$. Since $k > DH > h$ and
$k - h > 0$, we have: $a - b > k - h$ or $a + h > b + k$

In figure 2, assume CA > BC.

Choose point E on CA such that CE = CB. We have $b - a = EA > AF = h - k$. So ABC cannot be a scalene triangle. For $a + h = b + k$ to occur, $b = a$ and $h = k$, or CAB is isosceles with CA = CB.

Equilateral triangle CBA is also one of the conditions.

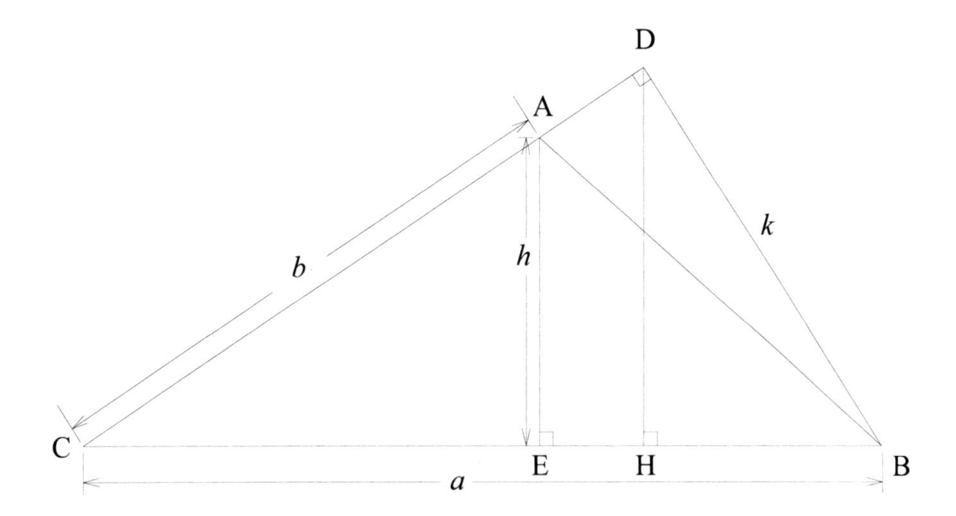

Figure 1

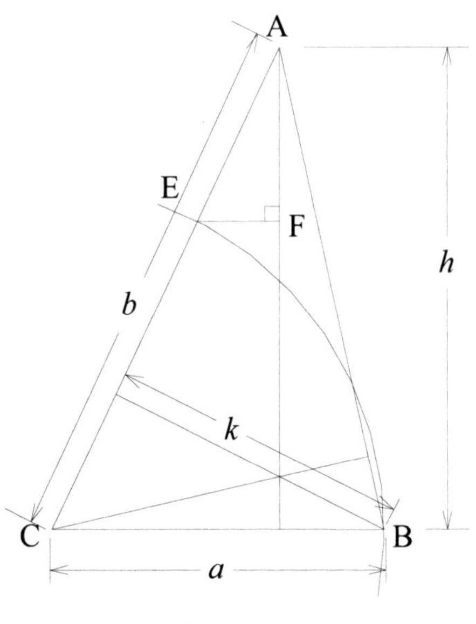

Figure 2

Problem 2 of the Canadian Mathematical Olympiad 1971

Let x and y be positive real numbers such that $x + y = 1$. Show that

$(1 + 1/x)(1 + 1/y) \geq 9$

Solution

Apply the AM-GM inequality, we have $x + y \geq 2\sqrt{xy}$ (i)

or $(x + y)^2 \geq 4xy$ or $1 \geq 4xy$,

or $xy \geq 4x^2y^2$, or $\sqrt{xy} \geq 2xy$, or $4\sqrt{xy} \geq 8xy$

Since $x + y = 1$, (i) can also be written as $1 \geq 2\sqrt{xy}$,

or $2 \geq 4\sqrt{xy} \geq 8xy$

or $x + y + 1 \geq 8xy$, or $(x + y + 1) / (xy) \geq 8$,

or $1/x + 1/y + 1/(xy) \geq 8$,

or $1 + 1/x + 1/y + 1/(xy) \geq 9$,

or $(1 + 1/x)(1 + 1/y) \geq 9$.

Problem 2 of the Canadian Mathematical Olympiad 1973

Find all the real numbers which satisfy the equation $|x+3| - |x - 1| = x + 1$. (Note: $|a| = a$ if $a \geq 0$; $|a| = -a$ if $a < 0$.)

Solution

For $x \in (-\infty, -3]$ (including point -3), the equation can be written as

$-x - 3 + x - 1 = x + 1$ or $x = -5$

For $x \in (-3, 1]$ (including point 1), the equation can be written as

$x + 3 + x - 1 = x + 1$ or $x = -1$

For $x \in (1, +\infty)$ (excluding point 1), the equation can be written as

$x + 3 - x + 1 = x + 1$ or $x = 3$

All the real numbers which satisfy the equation $|x+3| - |x - 1| = x + 1$ are -5, -1 and 3.

Problem 2 of the Canadian Mathematical Olympiad 1974

Let ABCD be a rectangle with BC = 3AB. Show that if P, Q are the points on side BC with BP = PQ = QC, then $\angle DBC + \angle DPC = \angle DQC$.

Solution

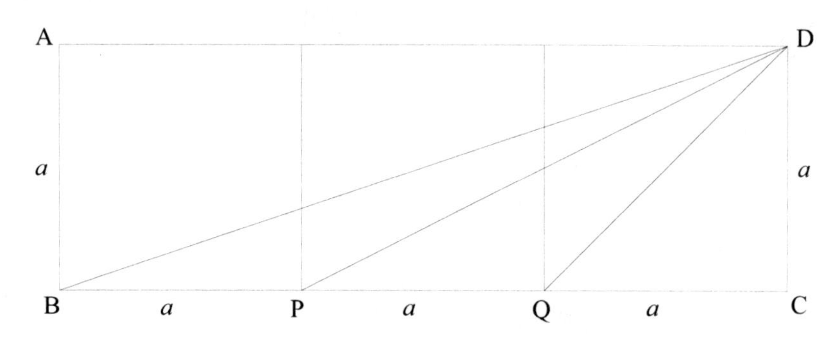

Apply the law of the cosine function for triangles PDQ and DBP

$PQ^2 = PD^2 + DQ^2 - 2\ PD \times DQ \cos\angle PDQ$ and
$PD^2 = BP^2 + BD^2 - 2\ BP \times BD \cos\angle DBP$ or

$a^2 = 7\ a^2 - 2\ a^2 \cos\angle PDQ \sqrt{10}$ and

$a^2 = 19\ a^2 - 6\ a^2 \cos\angle DBP \sqrt{10}$

Equating those two equations, we have

$$\cos\angle PDQ \sqrt{10} = -6 + 3 \cos\angle DBP\sqrt{10} \qquad \text{(i)}$$

but $\cos\angle DBP = BC / BD = 3 / \sqrt{10}$, we then have $\cos\angle PDQ \sqrt{10} = 3$ or $\cos\angle PDQ = 3/\sqrt{10} = \cos\angle DBP$ and since both $\angle DBP$ and $\angle PDQ$ are acute they must be equal $\angle DBP = \angle PDQ$

Now $\angle DQC = \angle DPC + \angle PDQ = \angle DBC + \angle DPC$.

Problem 2 of the Canadian Mathematical Olympiad 2010

Let A, B, P be three points on a circle. Prove that if a and b are the distances from P to the tangents at A and B and c is the distance from P to the chord AB, then $c^2 = ab$.

Solution

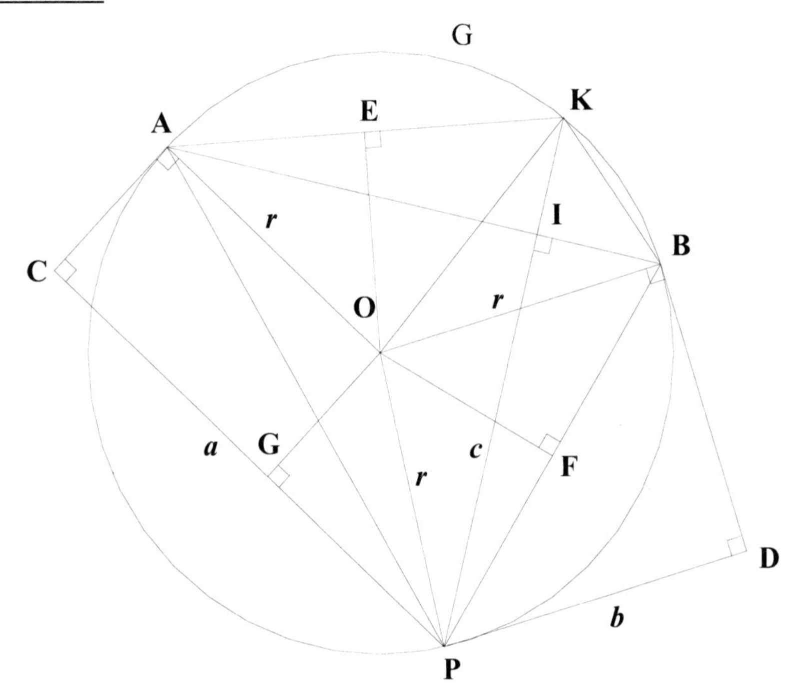

Let O be the center of the circumcircle Γ of triangle ABP, I be the foot of P to AB, PI = c. Extend PI to meet Γ at K.

Now let E and F be the midpoints of AK and BP, respectively; E and F are also the feet of O to AK and BP, respectively.

From P draw the perpendicular lines to meet the tangents at A and at B of circumcircle Γ at C and D, respectively, PC = a, PD = b. Let G be the foot of O to PC and r the radius of circle Γ.

Since AB ⊥ PK, angles subtending arcs AK plus BP is 90° or

$\angle AOE\ (\frac{1}{2}\angle AOK) + \angle POF\ (\frac{1}{2}\angle POB) = 90°$

but $\angle AOE + \angle EAO = 90°$ or $\angle EAO = \angle POF$

Combining with OA = OP = r, the triangles AOE and OPF are congruent and

AE = OF, and OE = PF.

We now have $AK^2 + PB^2 = (2AE)^2 + (2PF)^2 = 4AE^2 + 4OE^2 = 4r^2$. Moreover,

$AK^2 + PB^2 = AI^2 + KI^2 + PI^2 + BI^2$, or

$4r^2 = c^2 + AI^2 + BI^2 + KI^2$ (i)

Now the right triangle GOP gives us $OP^2 = OG^2 + GP^2$

or $r^2 = AC^2 + (a - r)^2$ or $a^2 = 2ar - AC^2$

or $2ar = AP^2$

Similarly, on the right side of the configuration $2br = BP^2$

Multiply the previous two equations, side by side

$4abr^2 = AP^2 \times BP^2$, or

$4abr^2 = (AI^2 + c^2)(BI^2 + c^2) = c4 + (AI^2 + BI^2)c^2 + AI^2 \times BI^2$

but $AI \times BI = c \times KI$ and now $4abr^2 = c4 + (AI^2 + BI^2 + KI^2)c^2$

Multiply both side of (i) by c^2, we have

$4c^2r^2 = c4 + (AI^2 + BI^2 + KI^2)c^2$

or $c^2 = ab$.

Problem 2 of the Ibero-American Mathematical Olympiad 1987

In a triangle ABC, M and N are the midpoints of the sides AC and AB respectively, and P is the point of intersection of BM and CN. Show that if it is possible to inscribe a circumference in the quadrilateral ANPM, then the triangle ABC is isosceles.

Solution

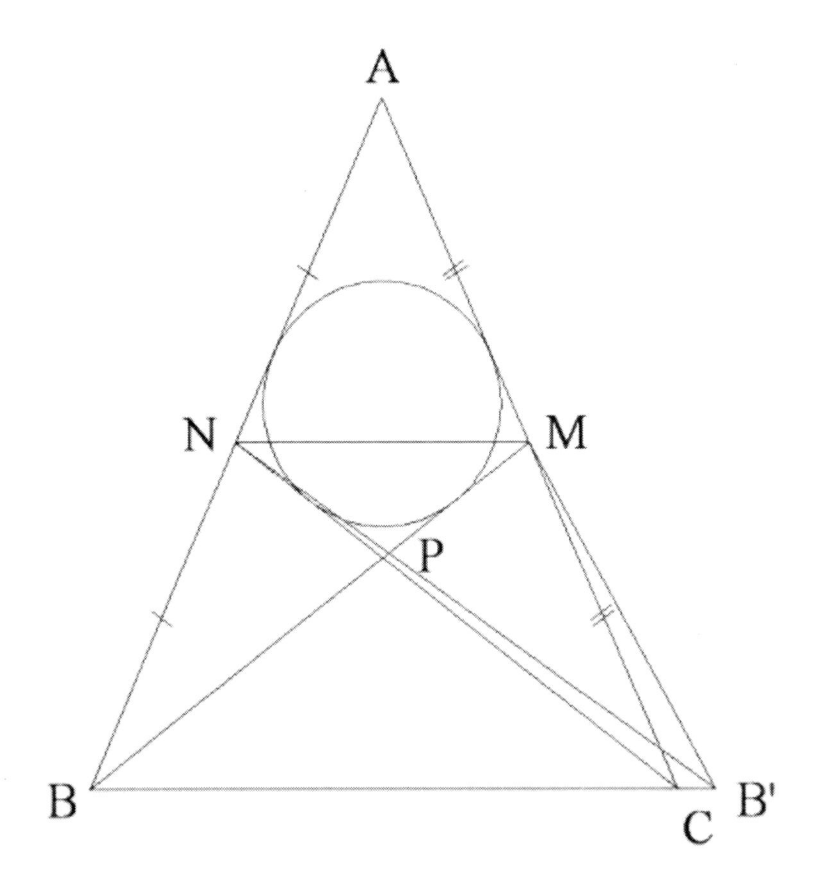

Since M and N are the midpoints of the sides of triangle ABC, area of the triangles ABM = area of triangle ACN = ½ area of triangle ABC. These two triangles also share the same incircle; therefore, their perimeters are also equal since the area of a triangle equals ½ the product of radius of incircle with the perimeter. We then have

$$AN + BN + BM + AM = AN + NC + MC + AM$$

or $\quad BN + BM = NC + MC \quad$ (i)

We know that NM || BC; let's pick point B' as mirror image of B via vertical line that goes through the center of NM and BC, and assume B' $\neq$ C.

If B' is on the right of C then $BN + BM = B'N + B'M > NC + MC$ since $B'N > CN$ and $B'M > MC$.

If B' is on the left of C, then $BN + BM < NC + MC$.

So to satisfy (i) we have to have B' $\equiv$ C (B' coincides with C), and therefore, BN = CM, and the triangle ABC is isosceles with AB = AC.

Problem 2 of the Ibero-American Mathematical Olympiad 1990

Let ABC a triangle, and let I be the center of the circumference inscrited and D, E, F its tangent points with BC, CA and AB respectively. Let P be the other point of intersection of the line AD with the circumference inscrited. If M is the mid point of EF, show that the four points P, I, M, D are either on the same circumference or they are collinear. (Exact wording from the exam text for easy Internet searching)

Solution

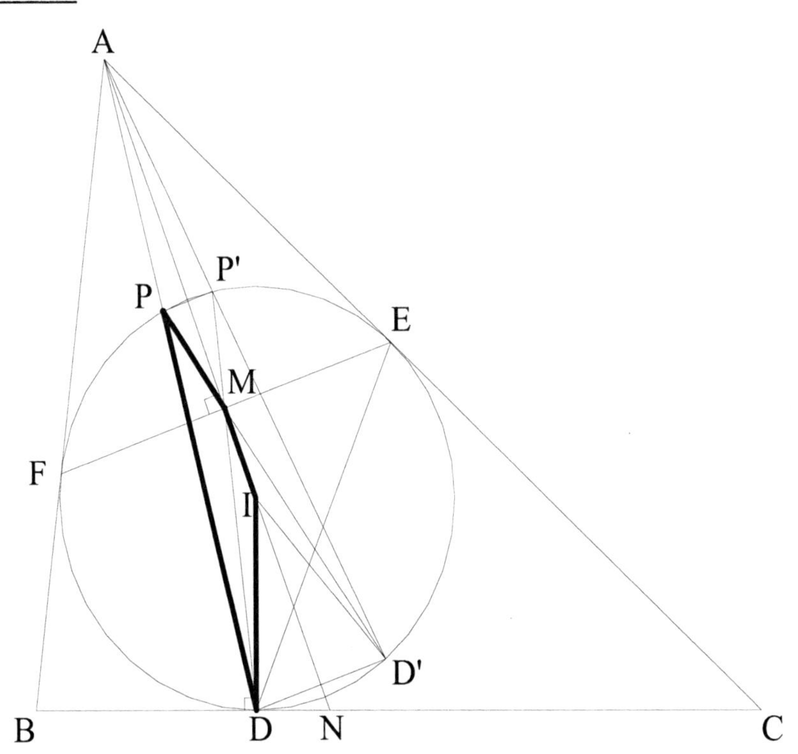

From D link and extend DM to meet the circle at P'. From P link and extend PM to meet the circle at D'. Since AN passes through the center of the circle and M is on AN, P' and D' are the mirror images of P and D via AN, respectively. And we also have

$PP' \parallel DD' \parallel EF \perp AN$.

That leads to ID = ID' and $\angle$DIN = $\angle$D'IN and since I is the center of the circle, we then have

$$\angle DIN = \angle DPD' = \angle DPM$$

$$\angle MID + \angle DIN = 180°, \text{ or}$$
$$\angle MID + \angle DPM = 180°, \text{ or}$$

PMID is a cyclic quadrilateral. The four points P, M, I and D will be on a straight line when the given triangle ABC is isosceles with AB = AC.

Problem 2 of the Ibero-American Mathematical Olympiad 1996

Let ABC be a triangle, D the midpoint of BC, and M be the midpoint of AD. The line BM intersects the side AC at point N. Show that AB is tangent to the circumcircle of the triangle NBC if and only if the following equality is true

$$\frac{BM}{MN} = \frac{BC^2}{BN^2}$$

Solution

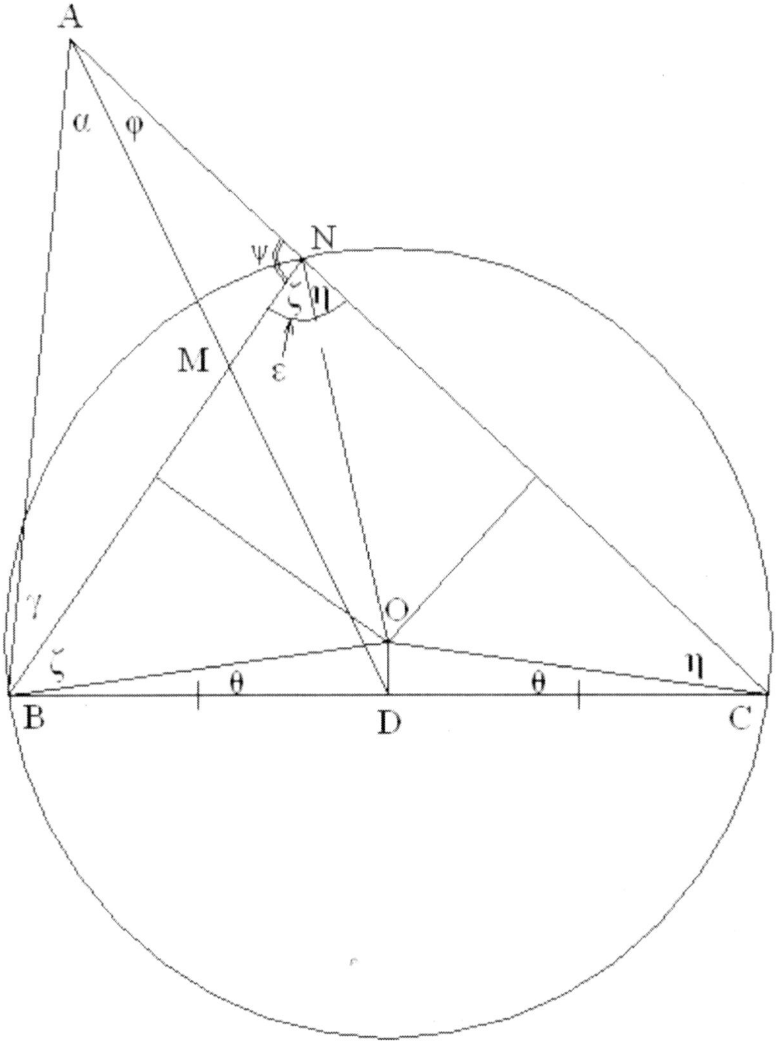

Let O be the center of the circumcircle of triangle NBC and let

$\angle BAM = \alpha$, $\angle MAN = \varphi$, $\angle ABM = \gamma$, $\angle ANM = \psi$, $\angle BNO = \angle NBO = \zeta$, $\angle ONC = \angle OCN = \eta$, $\angle OBC = \angle OCB = \theta$, and $\varepsilon = \angle BNC = \zeta + \eta = 180° - \psi$

***Given the condition that AB is tangent to the circumcircle of triangle NBC at B*, we have**

$\angle ABN = \gamma = \angle ACB = \eta + \theta$

Use the law of the sine function, we have BM/sinα = AM sinγ and MN/sinφ = AM/sinψ, or

$BM \sin\gamma / \sin\alpha = MN \sin\psi / \sin\varphi$, or
$BM / MN = \sin\psi \times \sin\alpha / \sin\varphi \times \sin\gamma$, and
$BC^2 / BN^2 = \sin^2\varepsilon / \sin^2\gamma$

so to prove $BM/MN = BC^2 / BN^2$, we have to prove

$$\frac{\sin\psi \times \sin\alpha}{\sin\varphi \times \sin\gamma} = \frac{\sin^2\varepsilon}{\sin^2\gamma} \qquad \text{(i)}$$

but $\sin\varepsilon = \sin(180° - \psi) = \sin\psi$, and (i) becomes

$$\frac{\sin\alpha}{\sin\varphi} = \frac{\sin\varepsilon}{\sin\gamma}, \quad \text{or} \quad \frac{\sin\alpha}{\sin\varphi} = \frac{\cos\theta}{\sin\gamma} \qquad \text{(ii)}$$

(because $\varepsilon = 90° - \theta$)

<u>Now let's prove it</u>
Again using the law of the sine function for triangle ABC, we have

$BD/ \sin\alpha = AD/ \sin(\gamma + \zeta + \theta)$ and $DC/ \sin\varphi = AD/ \sin\gamma$, or
$BD/ AD = \sin\alpha / \sin(\gamma + \zeta + \theta) = \sin\varphi / \sin\gamma$, or
$\sin\alpha / \cos\theta = \sin\varphi / \sin\gamma$ because $\gamma + \zeta = 90°$, and

$\sin(\gamma + \zeta + \theta) = \cos\theta$

or $\frac{\sin\alpha}{\sin\varphi} = \frac{\cos\theta}{\sin\gamma}$ which is equation (ii) above.

Now given the condition that $\frac{BM}{MN} = \frac{BC^2}{BN^2}$ ***and we will prove AB is tangent to the circumcircle of triangle NBC.***

$BM / MN = \sin\alpha \times \sin\psi / (\sin\varphi \times \sin\gamma)$, and
$BC^2/BN^2 = \sin^2\varepsilon / \sin^2(\eta + \theta)$

$\frac{BM}{MN} = \frac{BC^2}{BN^2}$ already gives us $\frac{\sin\alpha}{\sin\varphi} = \frac{\cos\theta}{\sin\gamma}$

The law of the sine function for triangle ABC gives us

$$\frac{\sin\alpha}{\sin\varphi} = \frac{\sin(\gamma + \zeta + \theta)}{\sin(\eta + \theta)} \quad \text{or} \quad \frac{\cos\theta}{\sin\gamma} = \frac{\sin(\gamma + \zeta + \theta)}{\sin(\eta + \theta)} \qquad \text{(iii)}$$

but $\eta + \theta = 90° - \zeta$ and $\sin(\eta + \theta) = \cos\zeta$.

Equation (iii) becomes

$\cos\zeta \times \cos\theta = \sin(\gamma + \zeta + \theta) \times \sin\gamma$, or

$\cos(\zeta + \theta) + \cos(\zeta - \theta) = -[\ \cos(2\gamma + \zeta + \theta) - \cos(\zeta + \theta)]$, or

$\cos(\zeta - \theta) = -\cos(2\gamma + \zeta + \theta)$, or

$\cos(\zeta - \theta) = \cos(180° - 2\gamma - \zeta - \theta)$, or

$\zeta - \theta = 180° - 2\gamma - \zeta - \theta$, or

$\gamma + \zeta = 90°$

or AB is perpendicular to OB and AB is tangent to the circle.

Problem 2 of the Ibero-American Mathematical Olympiad 1997

In a triangle ABC, it is drawn a circumference with center in the in−center I and that intersect twice each of the sides of the triangle: the segment BC on D and P (where D is nearer two B); the segment CA on E and Q (where E is nearer to C); and the segment AB on F and R (where F is nearer to A). Let S be the point of intersection of the diagonals of the quadrilateral EQFR. Let T be the point of intersection of the diagonals of the quadrilateral FRDP. Let U be the point of intersection of the diagonals of the quadrilateral DPEQ. Show that the circumferences circumscrites to the triangles FRT, DPU and EQS have a unique point in common.

Solution

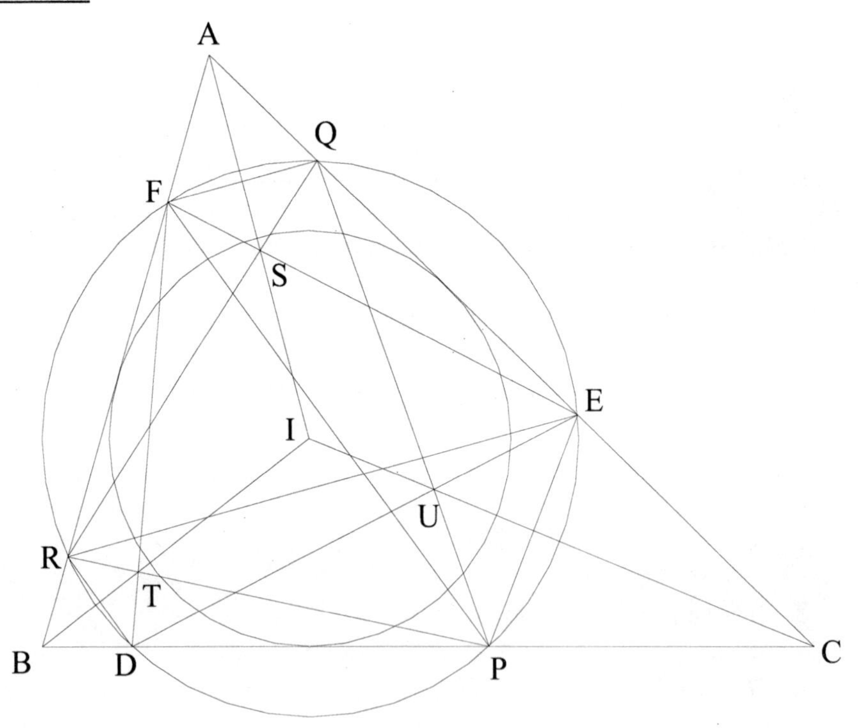

It is easily seen that BR = BD, and BI the angle bisector of $\angle B$ cuts RD in two equal segments.

Since we also have BF = BP, D and P are symmetrical images of R and F, respectively with respect to BI, and FRDP is a isosceles trapezoid, and T is on BI.

Similarly, EQFR and DPEQ are also isosceles trapezoids, and S and U are on segments AI and CI, respectively.

With I being the center of the larger circle, $\angle RID = 2\angle RFD$, or $\angle RFT = \angle RIT$ and FRTI is cyclic.

The same arguments apply to DPUI and EQSI. Therefore, the circumcircles of the triangles FRT, DPU and EQS have unique point I in common.

Problem 2 of the Ibero-American Mathematical Olympiad 1998

The circumference inscribed on the triangle ABC is tangent to the sides BC, CA and AB on the points D, E and F, respectively. AD intersect the circumference on the point Q. Show that the line EQ intersect the segment AF on its midpoint if and only if AC = BC.

Solution

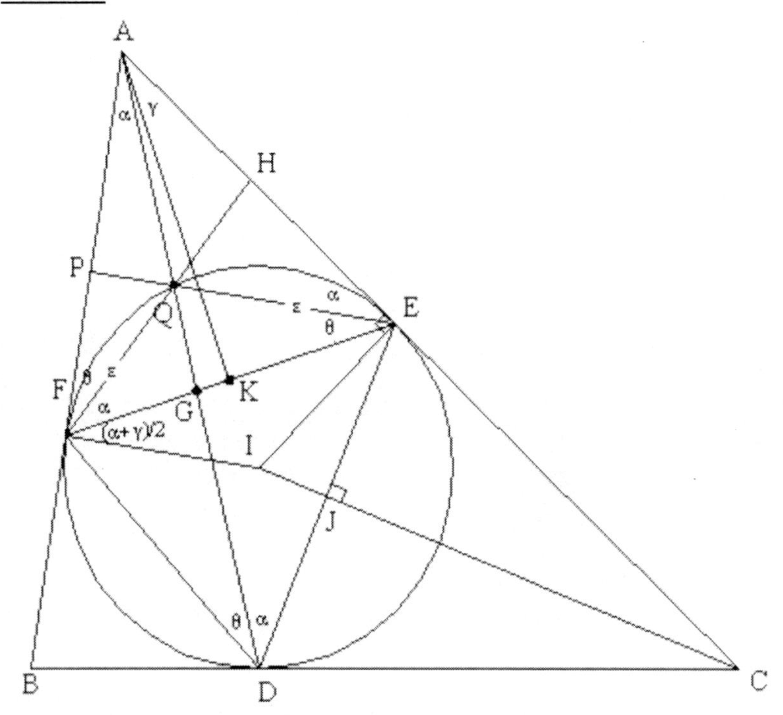

The case of AC = BC

Extend FQ to meet AC at H and EQ to meet AB at P. Let AD intercept EF at G.

Apply Ceva's theorem for the three lines AG, FH and EP, we have

$$\frac{AP}{PF} \times \frac{FG}{GE} \times \frac{EH}{AH} = 1$$

So, to prove point P, the intersection of EQ and AF, to be the midpoint of AF, we need to prove

$(FG / GE) \times (EH / AH) = 1$ (i)

Let $\angle FAG = \alpha$, $\angle GAE = \gamma$, $\angle AFQ = \theta$, $\angle AFE = \angle AEF = \varepsilon$, $\angle EFC = (\alpha+\gamma)/2$

We then also have
$\angle QDE = \alpha$ (since AB || ED) $= \angle QFE = \angle QEA$ (they subtend the same arc QE) and $\angle AFQ = \angle QEF = \angle QDF = \theta$ (subtending the same arc QF)

Apply the law of the sine function, we have

$FG / \sin\alpha = AG / \sin\varepsilon = GE / \sin\gamma$, or
$FG / GE = \sin\alpha / \sin\gamma$, and $EH = FH \sin\alpha / \sin\varepsilon$, and

$AH = FH \sin\theta / \sin(\alpha + \gamma)$

but $\alpha + \gamma = 180° - 2\varepsilon$ and $\sin(\alpha + \gamma) = \sin(180° - 2\varepsilon) = \sin 2\varepsilon = 2 \sin\varepsilon \cos\varepsilon = 2 \sin\varepsilon \cos(90° - \frac{1}{2} \angle A) = 2 \sin\varepsilon \sin(\frac{1}{2} \angle A) = 2 \sin\varepsilon \sin[(\alpha + \gamma)/2]$, or $AH = FH \sin\theta / \{ 2 \sin\varepsilon \sin[(\alpha + \gamma)/2] \}$

The equation (i) required to be proven becomes

$[2\sin^2\alpha \sin[(\alpha + \gamma)] / (2\sin\gamma \sin\theta) = 1$ (ii)

But in triangle AFD, $AF/\sin\theta = FD/\sin\alpha$, or $\sin\alpha/\sin\theta = FD/AF$ and in triangle AED, $\sin\alpha/\sin\gamma = AE/DE$ and in triangle AFK,

$\sin[(\alpha + \gamma)/2] = FK/AF$ and with $AE = AF$.

Equation (ii) then becomes

$FD \times EF = DE \times AF$ but $FD = EF$ or we need to prove
$EF^2 = DE \times AF$

Let's prove it

Again using the law of the sine function, in triangle AFK we have

FK $/\sin\angle A/2 = EF/(2\sin\angle A/2) = AF$ and in triangle FEJ, we have

$EJ / \sin\angle EFJ = EJ / \sin\angle A/2 = EF$, or

$EF^2 = (EJ/\sin\angle A/2) \times (AF \times 2\sin\angle A/2) = 2\,EJ \times AF = DE \times AF$

The case of AP = FP

Since AC is not yet equal to BC, we let $\angle QDE = \psi = \angle QFE = \angle QEA$ (they subtend the same arc QE)

We have $FG / GE = \sin\alpha / \sin\gamma$, and

$EH = FH \sin\psi / \sin\varepsilon$, and
$AH = FH \sin\theta / \{ 2\sin\varepsilon \sin[(\alpha + \gamma)/2] \}$, and

$[2\sin\alpha \sin\psi \sin[(\alpha + \gamma)]] / (2\sin\gamma \sin\theta) = 1$

which leads to $2\,EK \times FD = DE \times AF$, or

$FD \times EF = DE \times AF$, or

$FD = JE/\sin\angle A/2$ which only occurs when AC = BC.

Problem 2 of the Ibero-American Mathematical Olympiad 2001

The inscrite circumference of the triangle ABC has center at O and it is tangent to the sides BC, AC and AB at the points X, Y and Z, respectively. The lines BO and CO intersect the line Y Z at the points P and Q, respectively. Show that if the segments XP and XQ have the same length, then the triangle ABC is isosceles.

Solution

We have $\angle BOQ = \angle BCO + \angle OBC = \frac{1}{2}(180° - \angle A) =$
$\angle YZA = \angle ZOA$, and
$\angle OZQ = \angle ZAO$ (2 sides perpendicular to each other), or
$\angle BOQ + \angle BZQ = \angle YZA + 90° + \angle ZAO = 180°$
Therefore, BZQO is cyclic and $\angle BQO = \angle BZO = 90°$.
Similarly, $\angle CPO = 90°$ and
since $\angle BXO = \angle CXO = 90°$, BZQOX and CYPOX are both cyclic. We also note that BQPC is also cyclic.

Therefore, $\angle QXO = \angle QBO = \angle PCO = \angle PXO$ and triangles QXO and PXO are congruent which leads to OQ = OP and
$\angle QOX = \angle POX$, or
$\angle QBX = 180° - \angle QOX = 180° - \angle POX = \angle PCX$ (i)

Since $\angle OQP = \angle OPQ$ (OQ = OP) and $\angle OZY = \angle OYZ$ (OZ = OY = radius of the circle), triangles OZQ = triangle OYP or ZQ = YP. Furthermore, $\angle ZQX = 180° - \angle XQP = 180° - \angle XPQ =$ $\angle YPX$ and triangle ZQX = triangle YPX which leads to $\angle ZXQ$ $= \angle YXP$

Adding $\angle ZBQ$ to both sides of (i) $\angle QBX = \angle PCX$, we have

$\angle ZBX = \angle ZBQ + \angle QBX = \angle ZXQ + \angle QBX = \angle YXP + \angle PCX$
$= \angle YCP + \angle PCX = \angle YCX$

or AB = AC and the triangle ABC is isosceles.

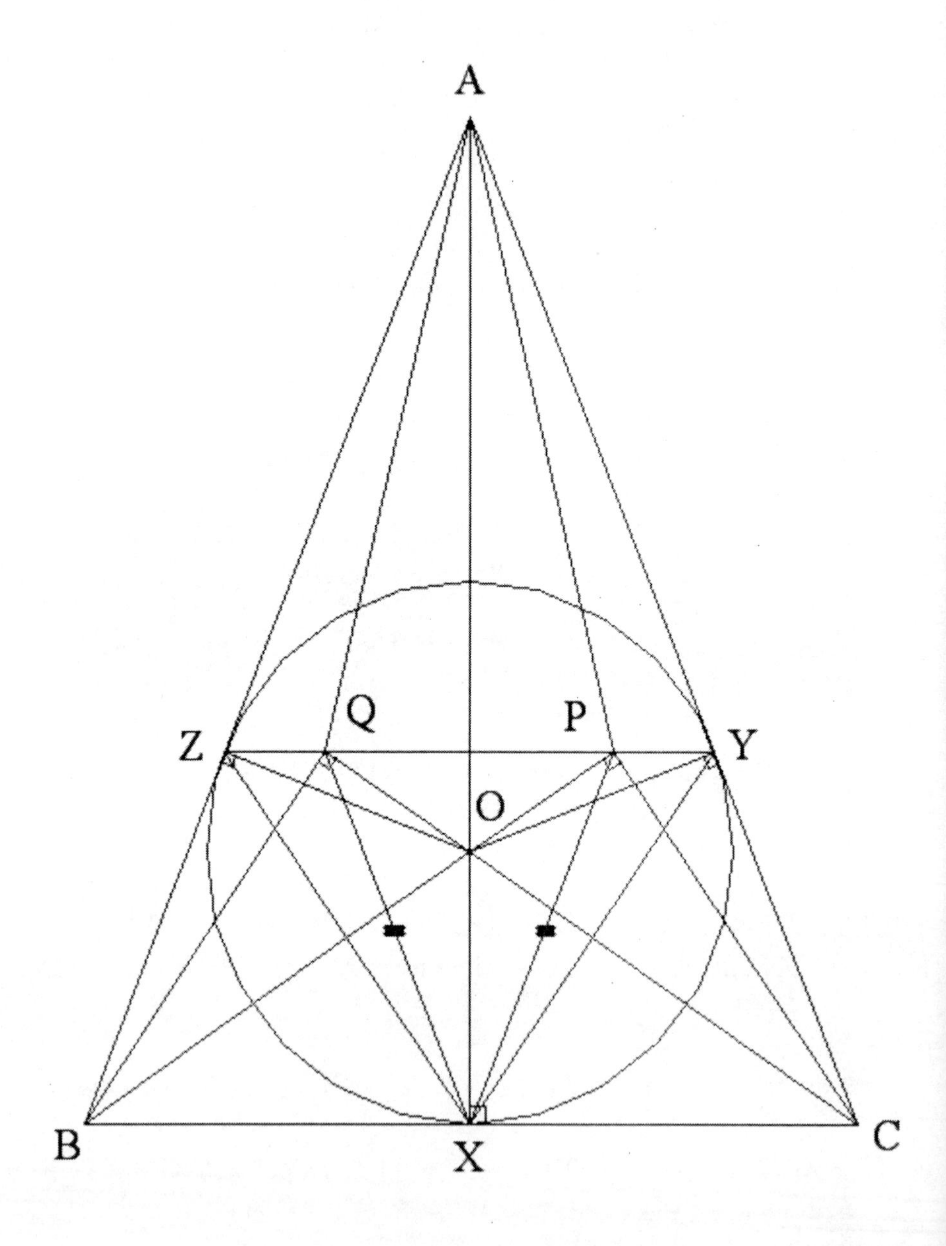
A
Q
P
Z
Y
O
B
X
C

Problem 2 of USA Stanford Mathematical Tournament 2006

Find the minimum value of $2x^2 + 2y^2 + 5z^2 - 2xy - 4yz - 4x - 2z + 15$ for real numbers x, y, z.

Solution

$2x^2 + 2y^2 + 5z^2 - 2xy - 4yz - 4x - 2z + 15 = x^2 + y^2 - 2xy + x^2 + 4 - 4x + z^2 - 2z + 1 + y^2 - 4yz + 4z^2 + 10 = (x - y)^2 + (x - 2)^2 + (z - 1)^2 + (y - 2z)^2 + 10$

The minimum values of all squares is zero, so the minimum value of the expression is 10.

Problem 2 of USA Mathematical Olympiad 1976

If A and B are fixed points on a given circle and XY is a variable diameter of the same circle, determine the locus of the point of intersection of lines AX and BY. You may assume that AB is not a diameter.

Solution

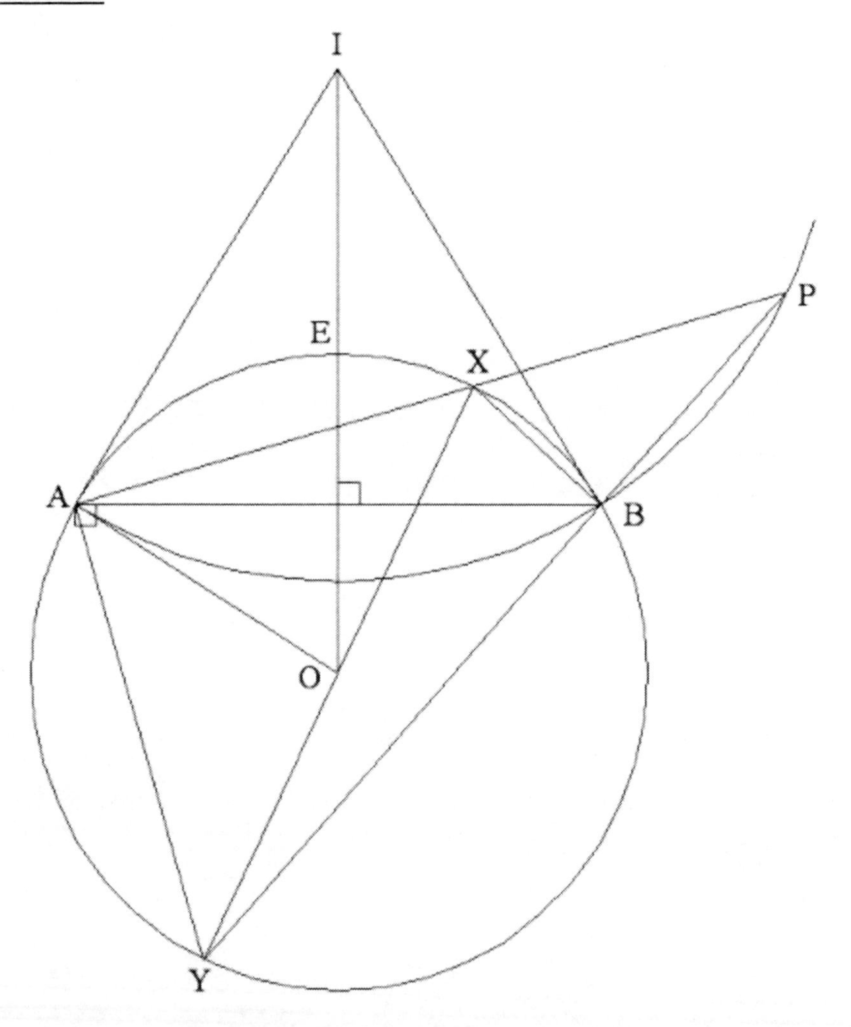

Draw the tangents of the circle at A and B to meet at I. Let IO intercept the circle at E between I and O.

We have $\angle AIO + \angle AOI = 90°$, but $\angle AOI = \angle AOE = \frac{1}{2}$ $\angle AOB = \angle AYB$ (O is center of circle and both $\angle AOB$ and $\angle AYB$ subtends arc AB).

It follows that $\angle AIO + \angle AYB = 90°$

On the other hand, since XY is the diameter of the circle $\angle XAY =$ $90° = \angle PAY$
or $\angle APY + \angle AYP = 90°$

Therefore, $\angle APY = \angle APB = \angle AIO = \frac{1}{2}\angle AIB$, or P is on the circle with fixed center I and fixed radius IA.

The locus is then a circle with center I and radius IA.

Problem 2 of USA Mathematical Olympiad 1993

Let ABCD be a convex quadrilateral such that diagonals AC and BD intersect at right angles, and let E be their intersection. Prove that the reflections of E across AB, BC, CD, DA are concyclic.

Solution

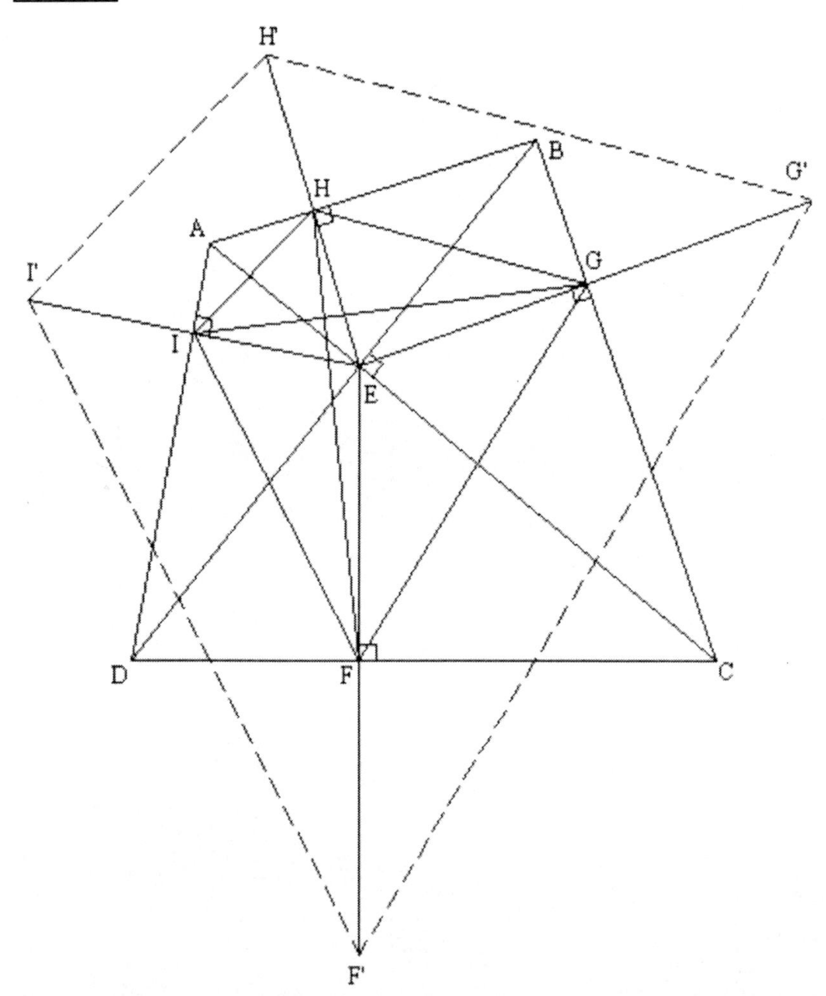

Let the feet from E to the four sides AB, BC, CD and DA be H, G, F and I as shown. Instead of proving the reflections of E

across AB, BC, CD, DA (H', G', F', and I') are concyclic we can prove H, G, F and I are concyclic because each foot is the midpoint of the distance from E to its reflection, and the two quadrilaterals are similar.

The four quadrilaterals EHBG, EGCF, EFDI and EIAH are cyclic since they have opposite right angles; we have

$\angle EHG = \angle EBG$, $\angle EFG = \angle ECG$, $\angle EFI = \angle EDI$,
and $\angle EHI = \angle EAI$

But since $AC \perp BD$, we have $\angle EBG + \angle ECG = 90°$ and $\angle EDI + \angle EAI = 90°$ or

$$\angle EHG + \angle EFG + \angle EFI + \angle EHI = 180°$$

or $\angle IHG + \angle IFG = 180°$ and H, G, F and I are concyclic.

<u>Further observation</u>

The problem below is derived from the one above.

Let ABCD be a convex quadrilateral such that diagonals AC and BD intersect at right angles, and let E be their intersection, and *C* be the circumcircle of triangle FGH where F, G and H are reflections of E across AB; BC and CD, respectively. Let I and J are points where DC intercepts the circle *C* and K is where the altitude line of triangle EDC from E intercept the circle *C*. Prove that E is the orthocenter of triangle KIJ.

Problem 2 of Vietnamese Regional Competition 1977

Compare $\frac{2^3+1}{2^3-1} \cdot \frac{3^3+1}{3^3-1} \cdot \cdot \cdot \frac{100^3+1}{100^3-1}$ with 3/2.

Solution

Rewrite the given expression $\frac{2^3+1}{2^3-1} \cdot \frac{3^3+1}{3^3-1} \cdot \cdot \cdot \frac{100^3+1}{100^3-1}$ as

$$(2^3+1). \frac{3^3+1}{2^3-1} \cdot \frac{4^3+1}{3^3-1} \cdot \cdot \cdot \frac{100^3+1}{99^3-1} \cdot \frac{1}{100^3-1} =$$

$$\frac{3^3+1}{2^3-1} \cdot \frac{4^3+1}{3^3-1} \cdot \cdot \cdot \frac{100^3+1}{99^3-1} \cdot \frac{2^3+1}{100^3-1} \qquad \text{(i)}$$

Note that $a^3 + b^3 = (a + b)(a^2 - ab + b^2)$ and $a^3 - b^3 = (a - b)(a^2 + ab + b^2)$

With b = 1, we have $a^3+1 = (a + 1)(a^2 - a + 1)$ and $(a - 1)^3 - 1 = (a-2)(a^2 - a + 1)$ and $\frac{a^3+1}{(a-1)^3-1} = \frac{a+1}{a-2}$.

We then write (i) as

$$\frac{3+1}{3-2} \cdot \frac{4+1}{4-2} \cdot \cdot \cdot \frac{100+1}{100-2} \cdot \frac{2^3+1}{100^3-1} = \frac{99\text{x}101}{2\text{x}3} \cdot \frac{2^3+1}{100^3-1} = 3/2 \times (9999/999999)\ .$$

We conclude that $\frac{2^3+1}{2^3-1} \cdot \frac{3^3+1}{3^3-1} \cdot \cdot \cdot \frac{100^3+1}{100^3-1} < 3/2$

Problem 3 of the Austrian Mathematical Olympiad 2004

In a trapezoid ABCD with circumcircle K the diagonals AC and BD are perpendicular. Two circles Ka and Kc are drawn whose diameters are AB and CD respectively.

Calculate the circumference and the area of the region that lies within the circumcircle K, but outside of the circles Ka and Kc.

Solution

Let (Ω) denote the area of shape Ω; also let AC intercept BD at I. Since AB || CD and ABCD is cyclic and also a trapezoid, it is an isosceles trapezoid. Triangles ABD and ABC are then congruent and $\angle ABD = \angle BAC$, $AD = BC$.

Let $a = AD = BC$, $b = AB$, $c = CD$. Because $AC \perp BD$, $\angle BAC = \angle ABD = \angle BDC = \angle ACD = 45°$.

Since $\angle BAC = 45°$ it subtends ¼ of the circumference of circle *K* or arc BC = arc AB = ¼ ($\pi \times$ diameter of *K*).

Similarly, $\angle BAC = \angle ABI = 45°$ and each subtends ¼ of the circumference of circle *Ka*, and $\angle IDC = \angle ICD = 45°$ and each subtends ¼ of the circumference of circle *Kc*.

Construct squares BCEF, AIBG and DICH with E and F on circle K, G on circle Ka and H on circle Kc, respectively. BE is the diameter of K. We then have

$$BE^2 = 2a^2 \text{ or } BE = a\sqrt{2}$$

The circumference in question = ½ circumference of K + ½ circumference of Ka + ½ circumference of Kc = ½ the sum of all

circumferences of three circles = $\frac{1}{2}(a\pi\sqrt{2} + b\pi + c\pi) = \pi(a\sqrt{2} + b + c)$

Apply Pythagorean's theorem, we have $a^2 = AI^2 + DI^2 = BI^2 + CI^2$ or

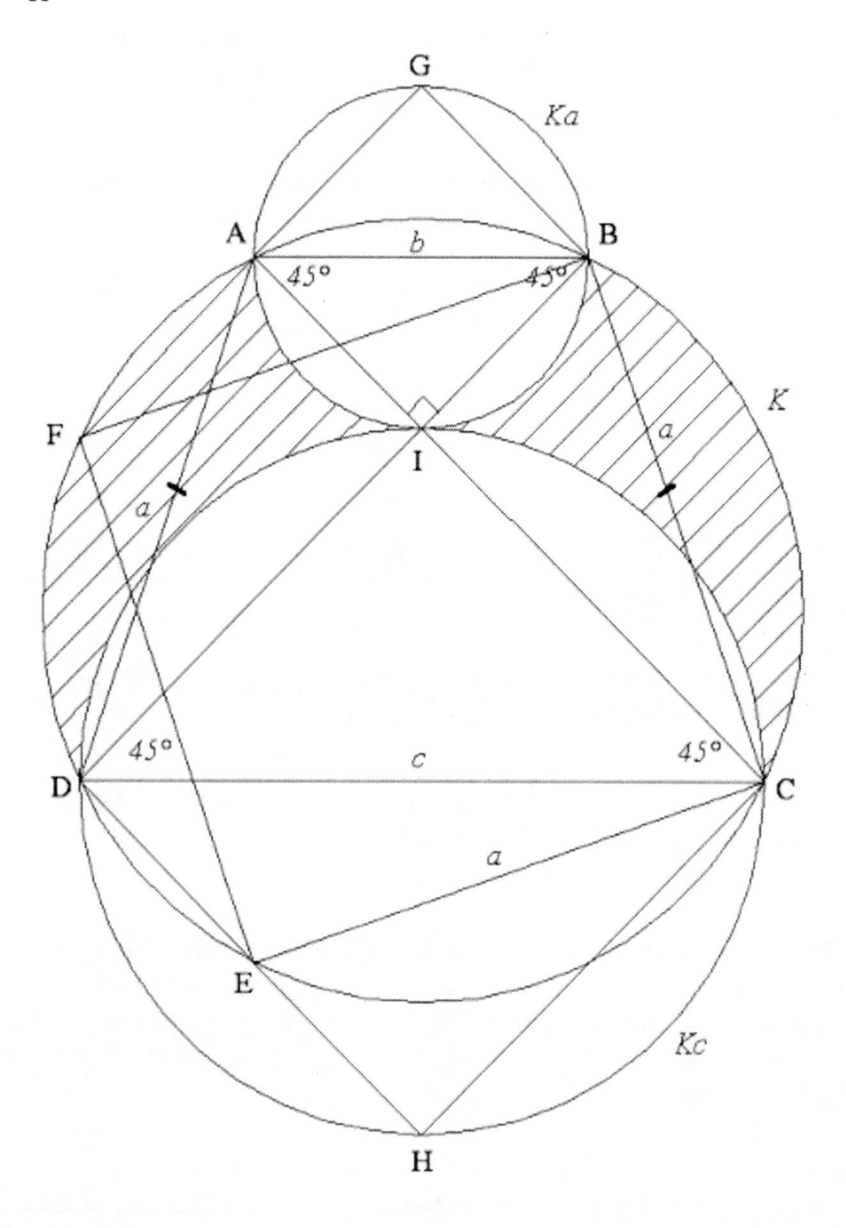

$2a^2 = b^2 + c^2$ or $a\sqrt{2} = \sqrt{b^2 + c^2}$

Therefore, the circumference is $\pi\,(b + c + \sqrt{b^2 + c^2}\,)$

Now the area of the region in question, let's call it A, has been shaded. One half of its area is equal to the area covered by BC and the smaller arc BD of circle K, this part equals to ¼ [the area of circle K – (FBCE)] + (BIC) – ¼ [the area of circle Ka – (AIBG)] – ¼ [the area of circle Ka – (CIDH)] = ¼ [(radius of K)$^2 \times \pi - a^2$] + ½ BI × CI – ¼ [(radius of Ka)$^2 \times \pi - BI^2$] – ¼ [(radius of Kc)^2 $\times \pi - CI^2$]

But we also have $2BI^2 = b^2$ and $2CI^2 = c^2$ or $BI = b/\sqrt{2}$ and

$CI = c/\sqrt{2}$

Therefore, half the area of the region is $\frac{1}{2}A = \frac{1}{4}(\pi a^2/2 - a^2) + bc/4 - \frac{1}{4}(\pi b^2/4 - b^2/2) - \frac{1}{4}(\pi c^2/4 - c^2/2) = [(\pi - 2)(2a^2 - b^2 - c^2) + 4bc]/16$

But again $2a^2 = b^2 + c^2$ and $\frac{1}{2}A = bc/4$ or $A = bc/2$

Problem 3 of the Austrian Mathematical Olympiad 2004

We are given a convex quadrilateral ABCD with $\angle$ADC = $\angle$BCD > 90°. Let E be the intersection of the line AC with the line parallel to AD through B and F be the intersection of the line BD with the line parallel to BC through A. Show that EF is parallel to CD.

Solution

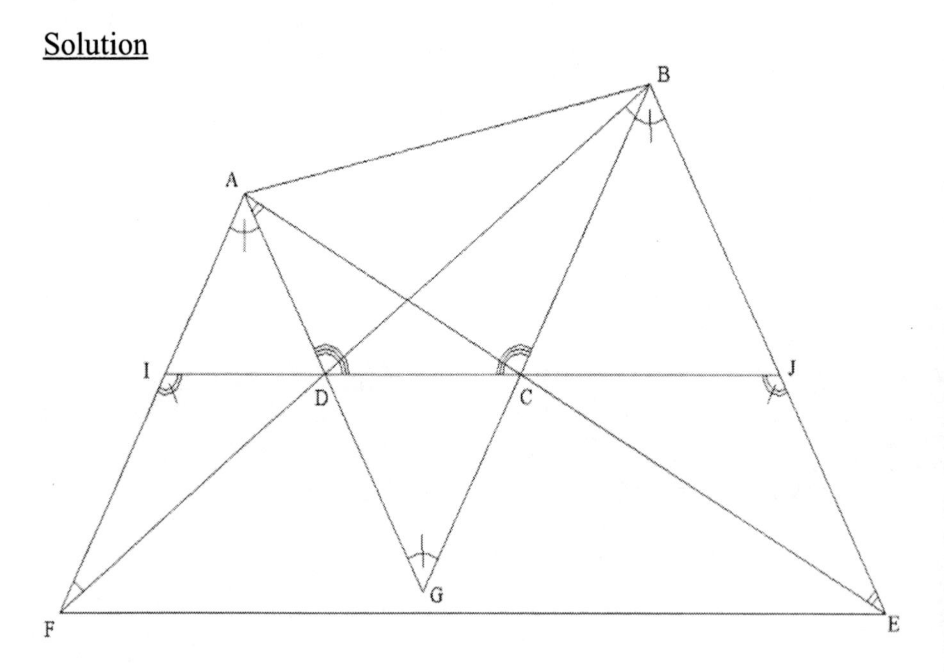

Extend both AD and BC to meet each other at G. Also extend DC to meet AF and BE at I and J, respectively.

Since AF || BC, AD || BE and $\angle$ADC = $\angle$BCD, we have $\angle$FAG = $\angle$AGB = $\angle$GBE, DG = CG, $\angle$ADI = $\angle$AID, $\angle$BCJ = $\angle$BJC. It follows that $\angle$FIJ = $\angle$EJI.

We also have $$\frac{AF}{CG + BC} = \frac{AD}{DG} = \frac{AF - AD}{BC} \qquad \text{(i)}$$

and $$\frac{BE}{AD + DG} = \frac{BC}{CG} = \frac{BE - BC}{AD} \qquad \text{(ii)}$$

Since DG = CG, (i) and (ii) give us AF – AD = BE – BC = AD × BC /DG.

But note that AD = AI, BC = BJ, we then haveIF = JE

Combining with ∠FIJ = ∠EJI, we conclude that IJEF is an isosceles trapezoid.

Problem 3 of the Austrian Mathematical Olympiad 2005

In an acute-angled triangle ABC two circles C1 and C2 are drawn whose diameters are the sides AC and BC. Let E be the foot of the altitude h*b* on AC and let F be the foot of the altitude h*a* on BC. Let L and N be the intersections of the line BE with the circle C1 (L on the line BE) and let K and M be the intersections of the line AF with the circle C2 (K on the line AF).

Show that KLMN is a cyclic quadrilateral.

Solution

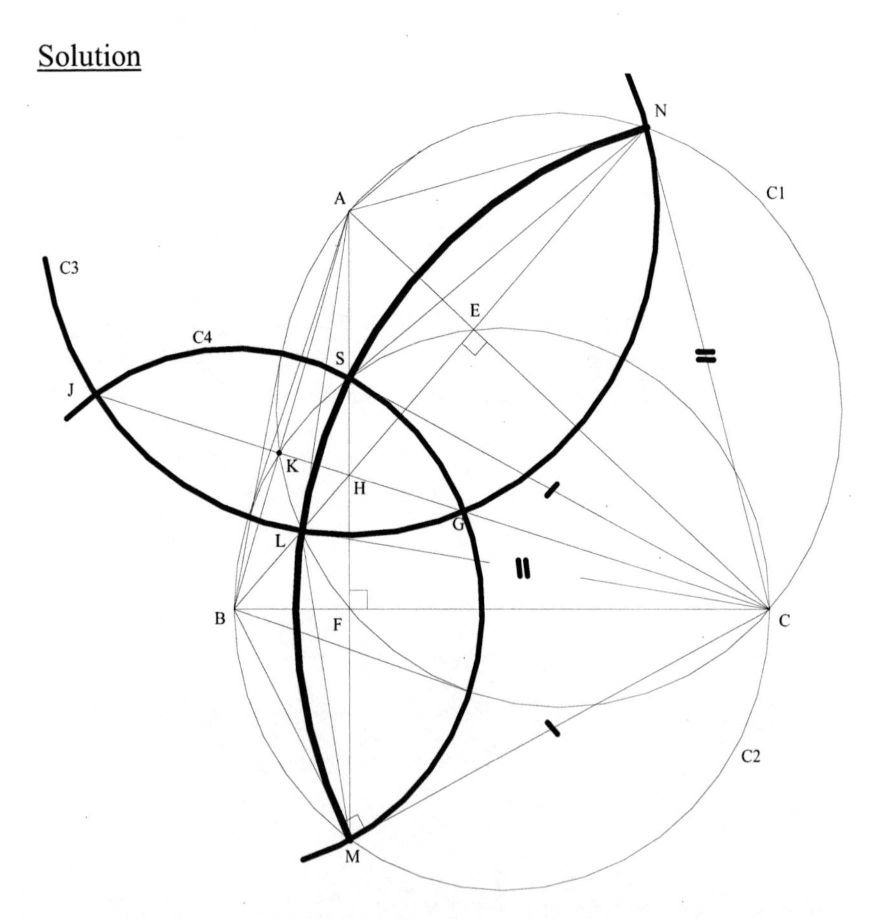

Let D be the foot of the altitude from C to AB. Since E is on the circle C2, BE $\perp$ AC and because AC is also the diameter of C1, AC is then the perpendicular bisector of LN. Therefore, CN = CL.

Similarly, since BC is the diameter of C2 and F is on circle C1, BC is perpendicular bisector of KM and CK = CM.

Now all we need to do is prove CM = CN so that the four points K, L, M, and N will lie on a circle with center at C and radius CN= CM = CK = CL.

Since the two triangles AFC and BEC are similar, we have

$CF/CE = CA/CB$ or $CF \times CB = CE \times CA$, or
$CF(CF + BF) = CE(CE + AE)$ or
$CF^2 + CF \times BF = CE^2 + CE \times AE$ (i)

But BMC is a right triangle at M and F its foot on BC, we have $CF \times BF = MF^2$, and similarly $CE \times AE = NE^2$

Now, rewrite (i) as $CF^2 + MF^2 = CE^2 + NE^2$, or $CM^2 = CN^2$ and we're done.

Further observation

Draw circle C3 with center A and radius AN = AL and circle C4 with center B and radius BM = BK. Let them intercept each other at J and G with G inside the circles. We can conclude that the four points D, H, G and C are collinear since $CM^2 = CN^2 = CG \times CJ$.

Problem 3 of the Austrian Mathematical Olympiad 2008

The line g is given, and on it lie the four points P, Q, R, and S (in this order from left to right).

Construct all squares ABCD with the following properties:
P lies on the line through A and D.
Q lies on the line through B and C.
R lies on the line through A and B.
S lies on the line through C and D.

Solution

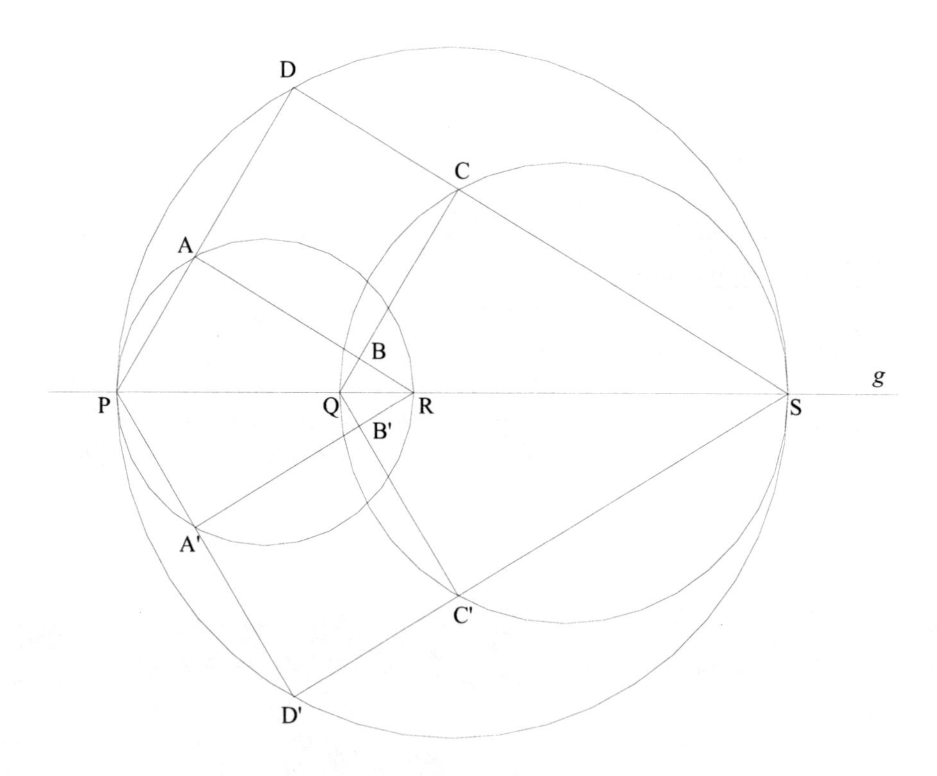

It is easily recognized that A must lie on the circle with diameter PR, D on circle with diameter PS, and C on circle with diameter QS. Let DS = d, CS = c, AP = a and

DP = b. For ABCD to be a square, AD || BC, DC || AB and $b - a = d - c$. We then have

$$d / PS = c / QS = (d - c) / (PS - QS) = CD / (PS - QS), \text{ and}$$

$$b / PS = a / PR = (b - a) / (PS - PR) = AD / (PS - PR)$$

Now CD = AD gives us:

$$d (PS - QS) / PS = b (PS - PR) / PS, \text{ or}$$

$$d = b (PS - PR)/ (PS - QS)$$

We also have $b^2 + d^2 = PS^2$

From those two equations, we have

$$b^2 + b^2(PS - PR)^2 / (PS - QS)^2 = PS^2 \text{ or}$$

$$b^2 [(PS - QS)^2 + (PS - PR)^2]/ (PS - QS)^2 = PS^2, \text{ or}$$

$$b = PS (PS - QS) /\sqrt{(PS - QS)^2 + (PS - PR)^2}$$

The square is then defined. Its mirror image A'B'C'D' across g is also a solution.

Problem 3 of Belarusian Mathematical Olympiad 2004

Find all pairs of integers (x, y) satisfying the equation
$y^2(x^2 + y^2 - 2xy - x - y) = (x + y)^2(x - y)$.

Solution

Expanding, eliminating and combining terms, we have
$y^2 (y - x)^2 = x^2 (y + x)$

Therefore, y + x must be a square of an integer. Let $y + x = n^2$ where n is an integer.

The above equation can be written as $y (y - x) = \pm nx$

Let's look at the case where $y (y - x) = nx$
Substituting $y = n^2 - x$ into the above equation, we have
$2x^2 - n(3n + 1)x + n^4 = 0$ now solving for x, we have

$x_{1\&2} = \frac{n}{4} [3n + 1 \pm \sqrt{n^2 + 6n + 1}\,]$ which requires $n^2 + 6n + 1$ to be a square of another integer. Let $n^2 + 6n + 1 = m^2$

Solving for n, we have $n_{1\&2} = -3 \pm \sqrt{m^2 + 8}$
Now $m^2 + 8$ must be a square or $m = \pm 1$
which makes n = 0 or n = -6

When n = 0, x = y = 0
When n = -6, x = 27, y = 9
x = 24, y = 12
And the other case $y (y - x) = - nx$

Similarly, the same procedure gives us $n_{1\&2} = 3 \pm \sqrt{m^2 + 8}$
and we end up having the same pairs of (x, y) as above.

Therefore, the three pairs of integers to satisfy the equation are

(x, y) = (0, 0), (27, 9) and (24, 12)

Problem 3 of the Canadian Mathematical Olympiad 1969

Let c be the length of the hypotenuse of a right angle triangle whose other two sides have lengths a and b. Prove that $a + b \leq c\sqrt{2}$. When does the equality hold?

Solution

Apply the AM-GM inequality for any non-negative real numbers a and b, we have

$(a + b)/2 \geq \sqrt{ab}$ and equality holds when $a = b$ or

$(a + b)^2 \geq 4ab$ or $a^2 + b^2 + 2ab \geq 4ab$, or

$$a^2 + b^2 \geq 2ab \qquad \text{(i)}$$

Since c is the hypotenuse of a right triangle and a and b are the other two sides

$$c^2 = a^2 + b^2 \text{ and (i) becomes } c^2 \geq 2ab \qquad \text{(ii)}$$

Now adding c^2 to the left and $a^2 + b^2$ to the right of (ii), we have

$2c^2 \geq (a + b)^2$ or

$a + b \leq c\sqrt{2}$ equality holds when $a = b$.

Problem 3 of the Canadian Mathematical Olympiad 1971

ABCD is a quadrilateral with AD = BC. If $\angle ADC$ is greater than $\angle BCD$, prove that $AC > BD$.

Solution

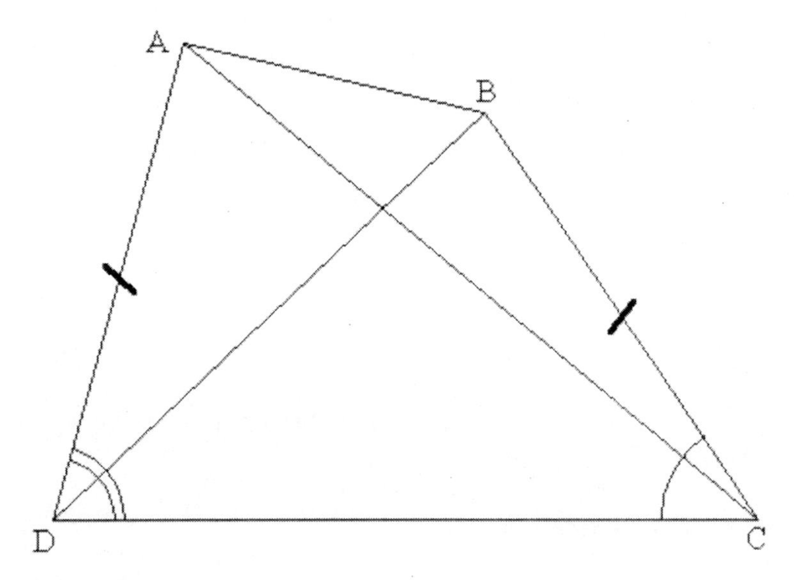

Apply the law of the cosine function:

$$AC^2 = DC^2 + AD^2 - 2\ DC \times AD \cos \angle ADC$$
$$BD^2 = DC^2 + BC^2 - 2\ DC \times BC \cos \angle BCD =$$
$$DC^2 + AD^2 - 2\ DC \times AD \cos \angle BCD$$

Since $\angle ADC > \angle BCD$, $\cos \angle ADC < \cos \angle BCD$ and $AC^2 > BD^2$ or $AC > BD$.

Problem 3 of the Canadian Mathematical Olympiad 1973

Prove that if p and p + 2 are both prime integers greater than 3, then 6 is a factor of p + 1.

Solution

Since p and p + 1 are prime integers, they are not divisible by 2 and we can express

$p = 2k + 1$ (k is an integer)
$p + 2 = 2k + 3$
or $p + 1 = 2(k + 1)$ or 2 is a factor of $p + 1$

and since they are not divisible by 3

$p = 3n + 1$ or $p = 3n + 2$ (n is an integer)

but if $p = 3n + 1$ then $p + 2 = 3(n + 1)$ which is divisible by 3

so the only option is $p = 3n + 2$ and $p + 2 = 3n + 4$

or $p + 1 = 3(n + 1)$ or 3 is also a factor of $p + 1$

Both 2 and 3 are factors of $p + 1$ then $2 \times 3 = 6$ is a factor of $p + 1$.

Problem 3 of the Canadian Mathematical Olympiad 1977

N is an integer whose representation in base b is 777. Find the smallest positive integer b for which N is the fourth power of an integer.

Solution

Let's write $N = 7b^2 + 7b + 7 = 7(b^2 + b + 1) = n^4$

Or $\quad b^2 + b + 1 = 7^3 m^4$ where m is a positive integer

The smallest positive integer b for which N is the fourth power of an integer is when $m = 1$ or $b^2 + b + 1 = 7^3$ or $b(b + 1) = 342$.

We have $18 \times 19 = 342$ or

$b = 18$ and then $N = 7(18^2 + 18 + 1) = 7^4$

Problem 3 of the Canadian Mathematical Olympiad 1978

Determine the largest real number z such that

$x + y + z = 5$
$xy + yz + xz = 3$

and x, y are also real.

Solution

From the top equation, $z = 5 - (x + y)$. To find the largest real number z we will find the numbers x and y such that $x + y$ is smallest.

From the bottom equation $z = \dfrac{3 - xy}{x + y}$ or $5 - (x + y) = \dfrac{3 - xy}{x + y}$

Rearrange this equation, we have $y^2 + (x - 5)y + x^2 - 5x + 3 = 0$ which has two roots as

$$y_{1\&2} = ½ (5 - x \pm \sqrt{-3x^2 + 10x + 13}\,)$$

Therefore, $x + y = x + ½ (5 - x \pm \sqrt{-3x^2 + 10x + 13}\,) = ½ (5 + x \pm$

$\sqrt{-3x^2 + 10x + 13}\,)$

And $x + y$ is at extreme when its derivative is equal to zero

$(5 + x \pm \sqrt{-3x^2 + 10x + 13}\,)' = 0$ or $1 \pm ½ \dfrac{1}{\sqrt{-3x^2 + 10x + 13}}$

$(-3x^2 + 10x + 13)' =$

$1 \pm ½ \dfrac{1}{\sqrt{-3x^2 + 10x + 13}}(-6x + 10) = 0$

Rearrange this equation and square both sides, we have
$3x^2 - 10x + 3 = 0$

This equation has solutions $x = 3$, and $x = \dfrac{1}{3}$

Substitute these x values to $x + y = \frac{1}{2}(5 + x \pm \sqrt{-3x^2 + 10x + 13})$

When $x = 3$, $x + y = 6$ and 2

When $x = \frac{1}{3}$, $x + y = \frac{2}{3}$ and $\frac{14}{3}$

Therefore, $x + y$ is minimum when $x + y = \frac{2}{3}$, and the largest value of z is $\frac{13}{3}$.

Problem 3 of the Canadian Mathematical Olympiad 1980

Among all triangles having (i) a fixed angle A and (ii) an inscribed circle of fixed radius r, determine which triangle has the least perimeter.

Solution

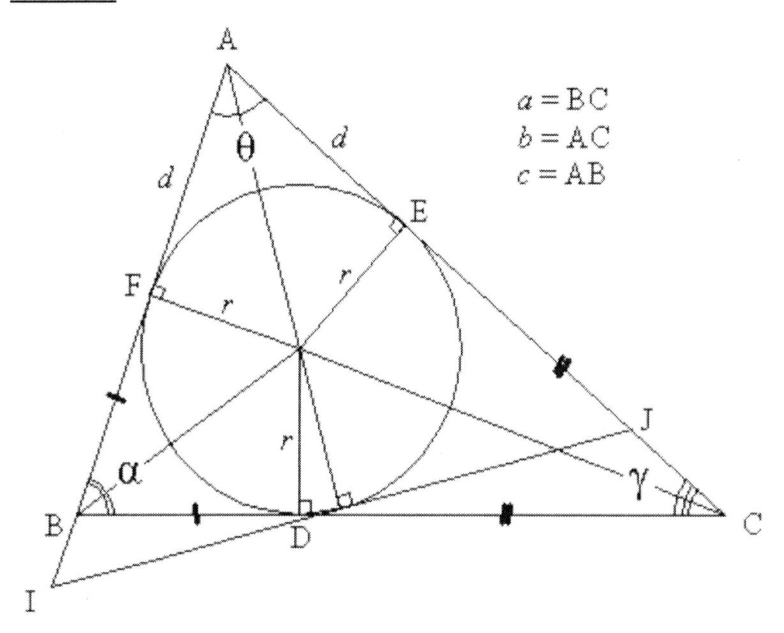

Let $\theta = \angle A$, $\alpha = \angle B$, $\gamma = \angle C$, r be the radius of the incircle, D, E and F are the points the incircle tangents with BC, CA and AB, respectively. Now let a = BC, b = AC, c = AB and d = AF = AE .

Note that BF = BD, CE = CD and we have

$a + b + c = 2d + 2BD + 2CD = 2d + 2(BD + CD) = 2d + 2a$ (i)

Since $\angle A$ and r are fixed, d is also fixed, and ***the minimum value of a + b + c is obtained when a is minimum***.

From (i) $a = b + c - 2d$ (ii)

Now apply the law of the sine function, we obtain
$a/\sin\theta = b/\sin\alpha = c/\sin\gamma$ or $c = b\sin\gamma / \sin\alpha$

Substituting them into (ii), we have

$a = b + b\sin\gamma / \sin\alpha - 2d = b\,[(\sin\alpha + \sin\gamma)/\sin\alpha] - 2d =$
$a\,[(\sin\alpha + \sin\gamma)/\sin\theta] - 2d$

or $\quad a = (2a/\sin\theta)\,[\cos \tfrac{1}{2}(\alpha - \gamma)\sin \tfrac{1}{2}(\alpha + \gamma)] - 2d$

or $\quad a = 2d\sin\theta / [2\cos \tfrac{1}{2}(\alpha - \gamma)\sin \tfrac{1}{2}(\alpha + \gamma) - \sin\theta]$

Since d, $\sin\theta$ and $\sin \tfrac{1}{2}(\alpha + \gamma)$ are all constants, a is minimum when $\cos \tfrac{1}{2}(\alpha - \gamma)$ is maximum or when it's equal to 1, or when $\alpha - \gamma = 0$ or $\alpha = \gamma$.

The triangle has the least perimeter when $\angle B = \angle C$ as in triangle AIJ shown on the graph.

Problem 3 of Canadian Mathematical Olympiad 1983

The area of a triangle is determined by the lengths of its sides. Is the volume of a tetrahedron determined by the areas of its faces?

Solution

There are two methods to prove this problem. One using a mathematical volume calculation, the other is easily proven using visual effect beyond doubt.

The first method:

The volume of a regular tetrahedron is given as

$V = A\mathbf{0}h/3 = a^3\sqrt{2}/12$ where a is the side of the tetrahedron. The area of an equilateral triangle which is the base of a tetrahedron depends on its side. So the volume of a tetrahedron depends on its area.

The second method:

Assume having two tetrahedrons ABCD and A'B'CD' with different side lengths a and b and $a > b$ as shown in the graph where they both are laying flat and being looked straight down (floor plan)

We can always make one of their vertices to coincide (vertices C in this case) and the sides A'D'C to lie completely on the plane of ADC and the same for B'D'C to lie on the plane of BDC. The volume of tetrahedron ABCD, therefore, completely covers that of tetrahedron A'B'CD'. So the area of the faces ABC and A'B'C of the tetrahedrons determine their volumes.

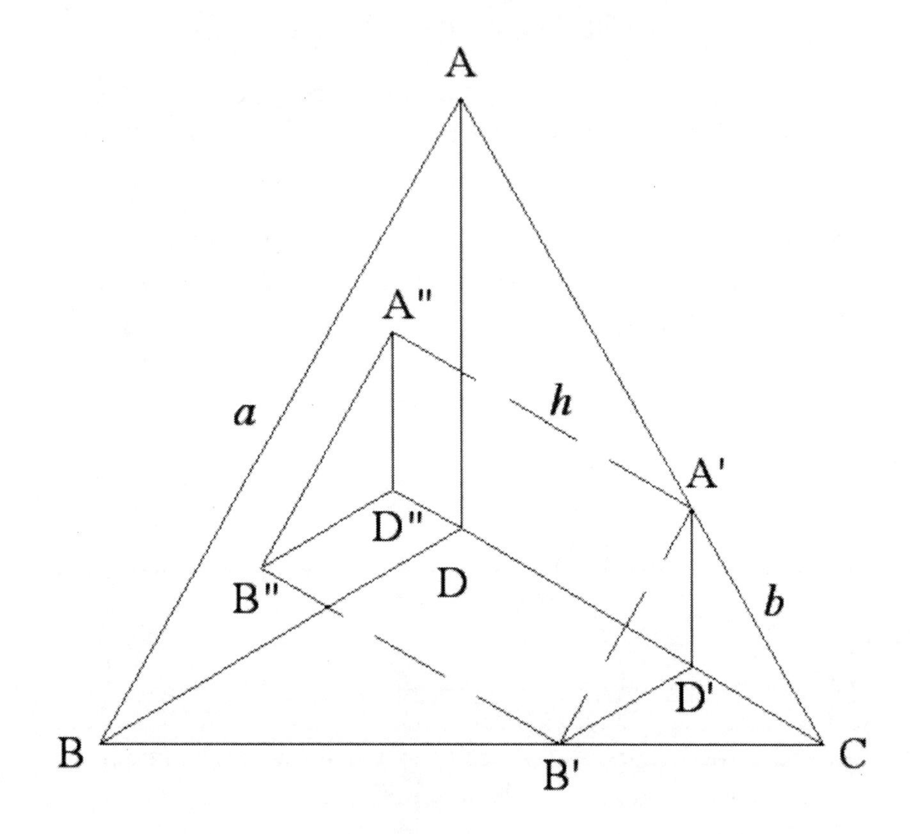

Or from A', B' and D' draw projections A", B", and D" to plane ABD, the volume of the tetrahedron ABCD is greater than that of A'B'CD' at least the volume of A"B"D"A'B'D' which is the area we can calculate without resorting to the formula above.

Problem 3 of Irish Mathematical Olympiad 2007

The point P is a fixed point on a circle and Q is a fixed point on a line. The point R is a variable point on the circle such that P, Q and R are not collinear. The circle through P, Q and R meets the line again at V. Show that the line VR passes through a fixed point.

Solution

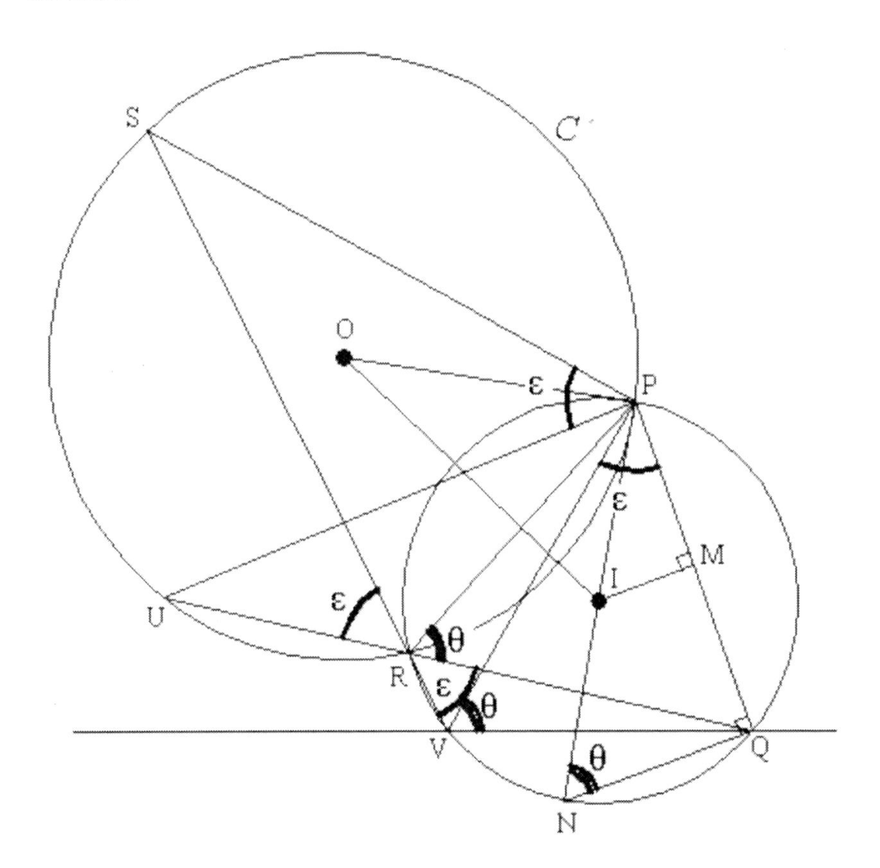

Let C be the circle where the fixed point P is on. Link QR and VR and extend them to intercept C at U and S, respectively. Let I and IN be the center and the diameter of the circumcircle of triangle PQR, respectively.

Now let $\varepsilon = \angle SPU$. We also have $\varepsilon = \angle SRU = \angle VRQ = \angle VPQ$ and let $\theta = \angle PRQ = \angle PVQ = \angle PNQ$.

We have $\angle SPQ = \angle UPQ + \angle SPU$ (i)

But O and I are centers of the two circles and P and R are their intersections, we then have

$\angle IOP = ½ \angle ROP = \angle PUQ$ and similarly $\angle OIP = \angle PQU$.
The two triangles OPI and UPQ are then similar and
$\angle OPI = \angle UPQ$

Equation (i) can now be written as

$\angle SPQ = \angle OPI + \varepsilon = \angle OPI + \angle VPQ = \angle OPI + 180° - \angle PQV - \angle PVQ = \angle OPI + 180° - \angle PQV - \angle PNQ = \angle OPI + 180° - \angle PQV - (90° - \angle NPQ) = \angle OPI + 180° - \angle PQV - 90° + \angle NPQ = \angle OPQ + 90° - \angle PQV$

Since both angles $\angle OPQ$ and $\angle PQV$ are constants, $\angle SPQ$ is then constant and VR passes through a fixed point.

Problem 3 of Middle European Mathematical Olympiad 2009

Let ABCD be a convex quadrilateral such that AB and CD are not parallel and AB = CD. The midpoints of the diagonals AC and BD are E and F, respectively. The line EF meets segments AB and CD at G and H, respectively. Show that $\angle AGH = \angle DHG$.

Solution

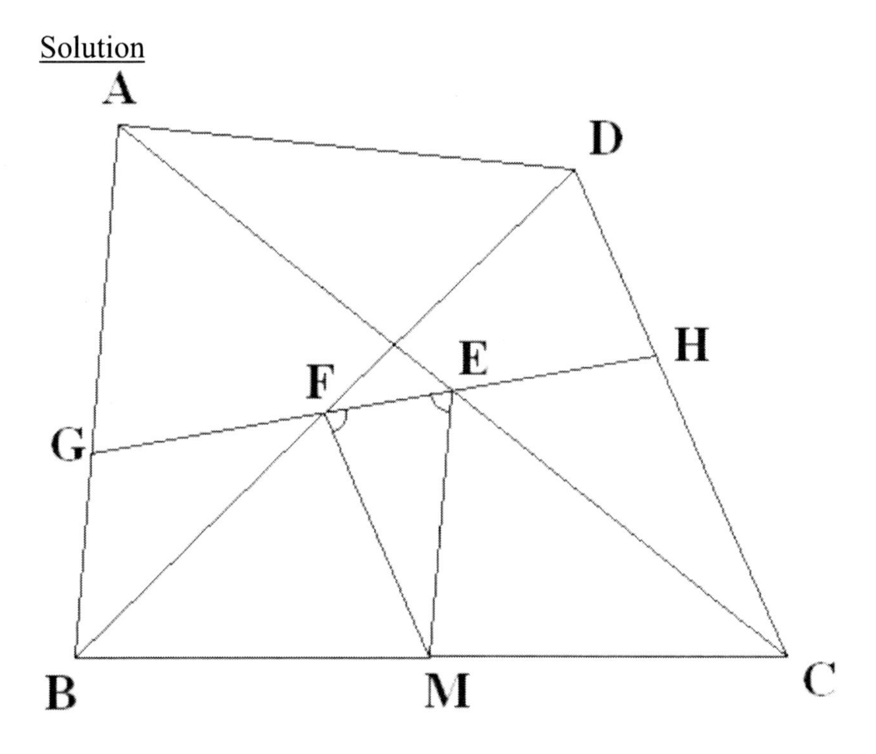

Let M be the midpoint of BC. We have EM = ½ AB, FM = ½ DC or FM = EM and $\angle EFM = \angle FEM$.

We also have EM || AB or $\angle AGH = \angle FEM$
FM || DC or $\angle DHG = \angle EFM$

Therefore, $\angle AGH = \angle DHG$

Problem 3 of the British Mathematical Olympiad 2005

Let ABC be a triangle with AC > AB. The point X lies on the side BA extended through A, and the point Y lies on the side CA in such a way that BX = CA and CY = BA. The line XY meets the perpendicular bisector of side BC at P. Show that
$\angle BPC + \angle BAC = 180°$.

Solution

Since BX = AC and AB = YC, we have AX = AY. Let $\angle BXY = \alpha$, we also have

$\angle AXY = \angle AYX = \angle PYC = \alpha$. Now let $\beta = \angle BAY$,
$\beta = \angle AXY + \angle AYX = 2\alpha$

From B draw a segment BJ such that BJ ||AY and BJ = AY. ABJY is a parallelogram and

$AB = YJ = YC$ and $\angle BAC = \angle JYC = 2\alpha$, or
$2\alpha = \angle PYC + \angle JYP = \alpha + \angle JYP$ or $\angle JYP = \alpha$

Now link JC and extend XY all the way to meet JC at N. Triangle JYN = triangle CYN.

Since they share YN, YJ = YC and $\angle JYN = \angle CYN$. Therefore, $YN \perp JC$ and PB = PC = PJ or P is the circumcenter of triangle BCJ.

Now draw the circumcircle of triangle BCJ, extend CA to meet the circle at I. Since P is the circumcenter, $\angle BPC = 2\angle BIC$ because both angles subtend the same arc BC. And since BJ || IC, BI = JC, we have JI = BC and $\angle BIC = \angle JCI = \delta$ as shown.

These two equations $\angle BPC = 2\angle BIC$ and $\angle BIC = \angle JCI = \delta$ give us $\angle BPC = 2\angle JCI = 2\delta$.

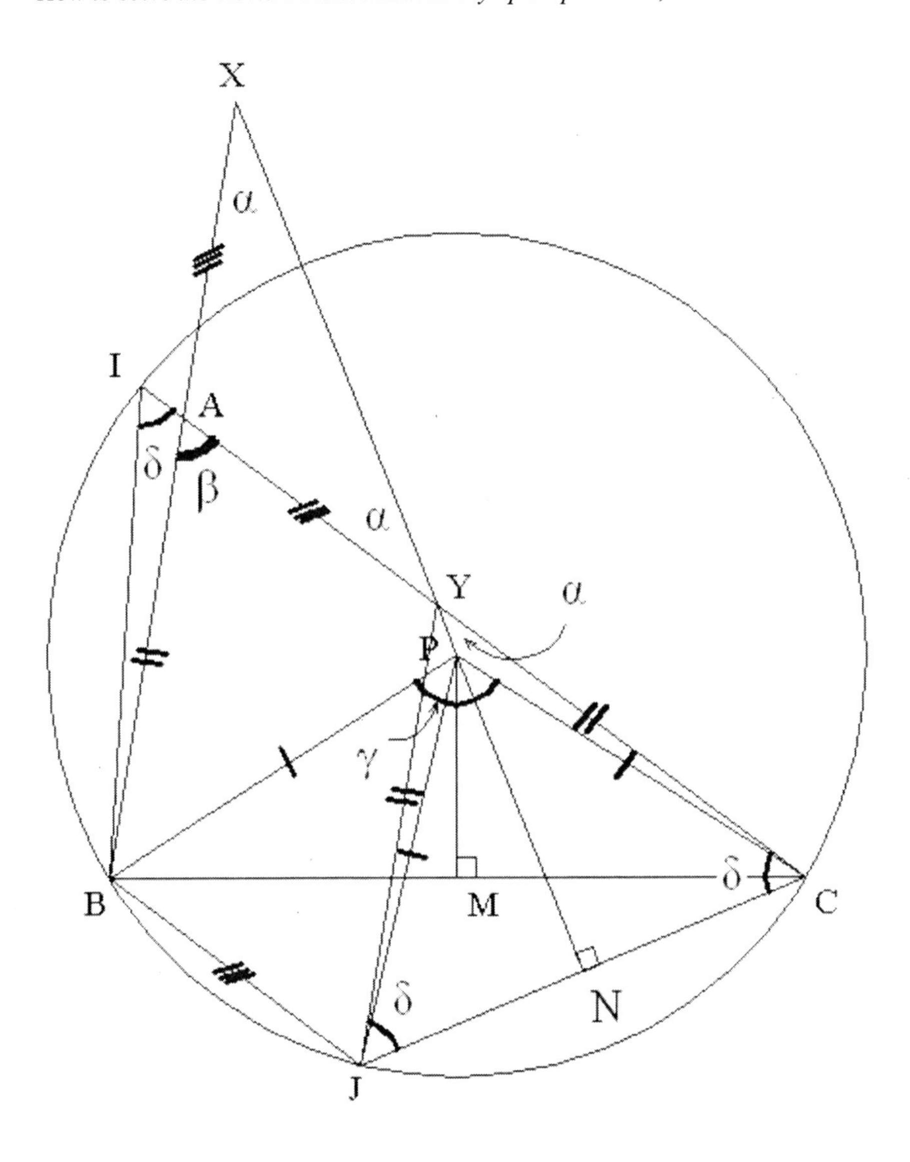

But $\quad 2\delta = \angle JCI + \angle YJC = \angle AYJ.$

Since ABJY is a parallelogram, $\angle AYJ + \angle BAC = 180°$, or

$2\delta + \angle BAC = 180°$ or $\angle BPC + \angle BAC = 180°$.

Problem 3 of the British Mathematical Olympiad 2006

Let ABC be an acute-angled triangle with AB > AC and ∠BAC = 60°. Denote the circumcenter by O and the orthocenter by H and let OH meet AB at P and AC at Q. Prove that PO = HQ.

Solution

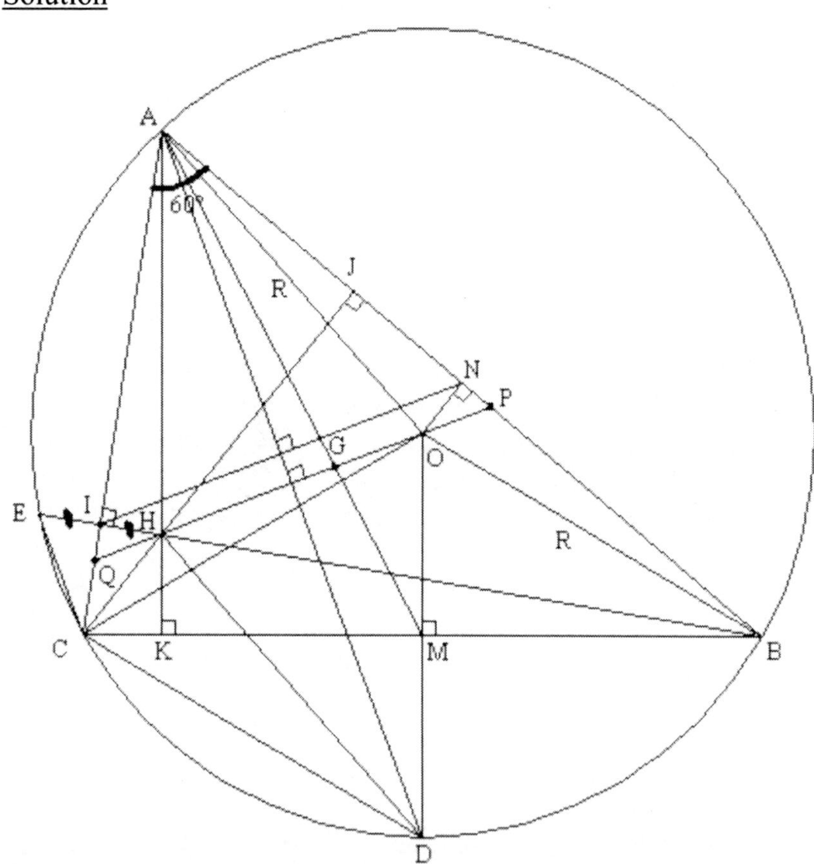

Let R be the radius of the circle, M and N the midpoints of BC and AB, respectively.
Extend OM to meet the circle at D. Extend CH to meet AB at J, BH to meet AC at I and the circle at E.

Since ∠BAC = 60°, ∠ABE = ACJ = ∠ACE (subtends arc AE) = ∠CAD (arc CD = ½ arc CB) = 30° or CE || AD

We also have $\angle COD = 2\angle CAD = 60°$ and OC = OD = R make OCD an equilateral triangle and since CM $\perp$ OD, we have $\angle OCB = \angle DCM = \angle OBC = 30°$

$\angle ABE = \angle ABO + \angle OBE = \angle OBC = \angle CBE + \angle OBE = 30°$
or
$\angle CBE = \angle ABO = \angle OAB$

Combining with $\angle CBE = \angle CAK = \angle IAH$, we have
$\angle OAB = \angle IAH$

Since AIB is a right triangle and N is the midpoint of AB, we have AN = NB = NI, and with $\angle BAC = 60°$, triangle ANI is equilateral and AI = AN

Now two right triangles AIH and ANO are congruent since all their corresponding angles are equal and AI = AN.

Therefore, AH = AO = R and since AH || OD it makes AODH a rhombus and the diagonal lines AD $\perp$ HO.

Also since $\angle AIN = 60°$ and $\angle IAD = 30°$ make AD $\perp$ IN we now have IN || HO, or $\angle ANI = \angle APQ = \angle AIN = \angle AQP = 60°$

Now the two triangles AHQ and AOP have all their corresponding angles equaled to one another and its sides AH = AO and are congruent.

Therefore, PO = HQ.

Problem 3 of Romanian Mathematical Olympiad 2006

In the acute-angle triangle ABC we have $\angle ACB = 45°$. The points A1 and B1 are the feet of the altitudes from A and B, respectively. H is the orthocenter of the triangle. We consider the points D and E on the segments AA1 and BC such that A1D = A1E = A1B1.
Prove that

a) $A1B1 = \sqrt{(A1B^2 + A1C^2)/2}$
b) CH = DE

Solution

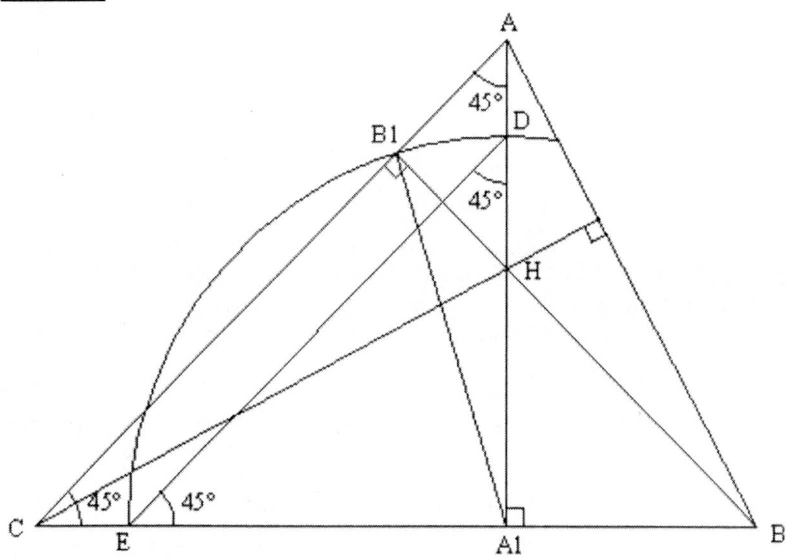

a) Since $\angle ACB = 45°$, $\angle CAA1 = 45°$, A1C = A1A and A1D = A1E causes $\angle A1ED = \angle A1DE = 45°$.

$A1B^2 + A1C^2 = AB^2 - AA1^2 + AC^2 - AA1^2 =$
$AB^2 + AC^2 - 2AA1^2$ (i)

But $2AA1^2 = AA1^2 + A1C^2 = AC^2$

and (i) becomes $A1B^2 + A1C^2 = AB^2$ (ii)

Since $\angle AB1B = \angle AA1B = 90°$, AB1A1B is cyclic, and the two triangles HB1A1 and HAB are similar, we have

$A1B1/AB = HB1/HA$

Furthermore, the three triangles A1ED, B1AH and A1CA are also similar, we have

$A1B1/DE = A1D/DE = AA1/AC = HB1/HA$

Therefore, $A1B1/AB = A1B1/DE$ or

$AB = DE$

Equation (ii) now becomes $A1B^2 + A1C^2 = DE^2$

We also have $2A1E^2 = A1E^2 + A1D^2 = DE^2$ or

$A1B^2 + A1C^2 = 2A1E^2 = 2A1B1^2$

Therefore, $A1B1 = \sqrt{(A1B^2 + A1C^2)/2}$

b) We have $A1H = A1B$ (right isoceles triangle A1HB), and $\angle HCA1 = \angle BAA1$ (sides perpendicular), $\angle CA1H = \angle AA1B = 90°$, the two triangles CHA1 and ABA1 are congruent and $CH = AB = DE$ ($AB = DE$ from part a)

Problem 3 of the Vietnam Mathematical Olympiad 1989

A square ABCD of side length 2 is given on a plane. The segment AB is moved continuously towards CD until A and B coincide with C and D, respectively. Let S be the area of the region formed by the segment AB while moving. Prove that AB can be moved in such a way that $S < \frac{5\pi}{6}$

Solution

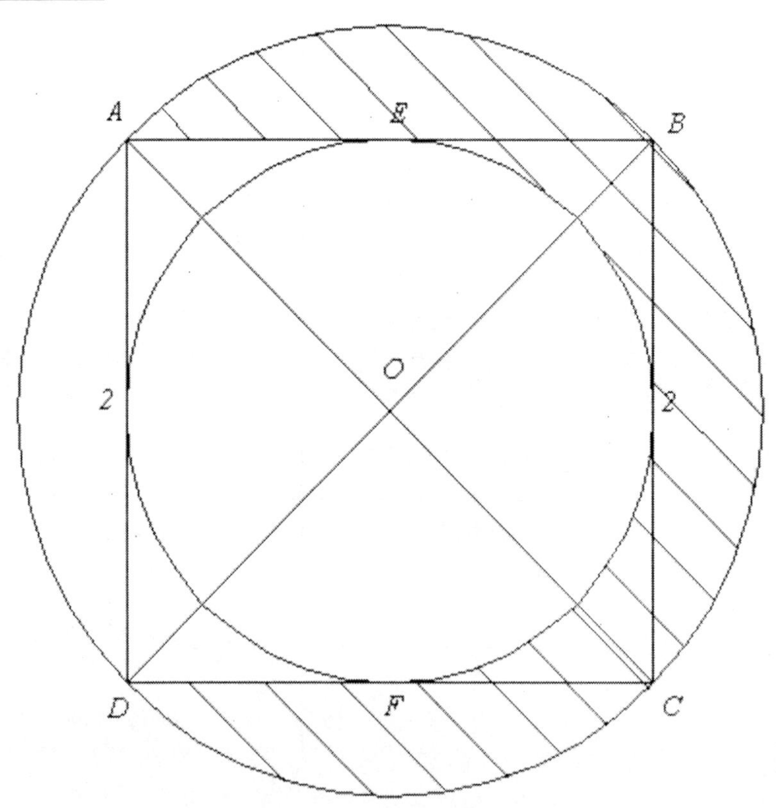

Let E and F be the midpoints of AB and CD, respectively and O the intersection of the diagonals of ABCD.

Let's move AB by rotating it clockwise around the center O. The area of the region formed by the segment AB while moving will be the shaded area as shown. If a vertical line is drawn to pass through the two midpoints E and F, one can easily see that this area equals ½ the area bounded between the two circles plus ¼ the area bounded between the outer circle and the square.

$S = \frac{1}{2}\pi(OA^2 - OE^2) + \frac{1}{4}(\pi \times OA^2 - AB^2) = \frac{1}{2}\pi(\sqrt{2}^{\,2} - 1^2) + \frac{1}{4}(\pi\sqrt{2}^{\,2} - 2^2) = \frac{1}{2}\pi + \frac{1}{4}(2\pi - 4) = \pi - 1 < \frac{5\pi}{6}$

Problem 4 of the Austrian Mathematical Olympiad 2002

We are given three mutually distinct points A, C and P in the plane. A and C are opposite corners of a parallelogram ABCD, the point P lies on the bisector of the angle DAB, and the angle APD is a right angle. Construct all possible parallelograms ABCD that satisfy these conditions.

Solution

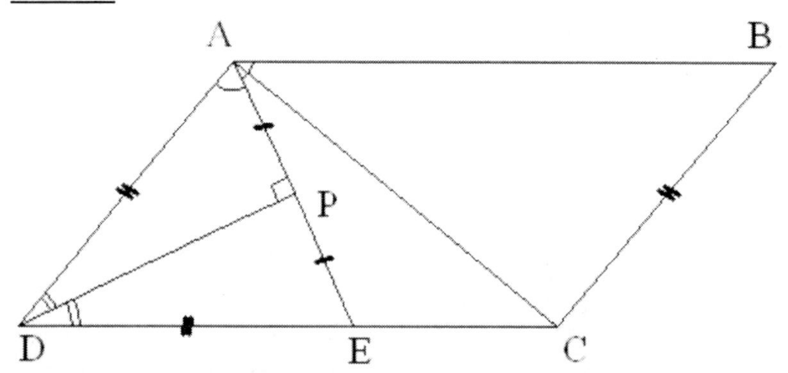

We know that point D is on the line Px perpendicular to AP at P.

Since AP is the bisector of $\angle$DAB, $\angle$DAP = $\angle$BAP. The sum of two consecutive angles of a parallelogram is 180°, or
$\angle$BAD + $\angle$ADC = 180°   or   2$\angle$DAP + $\angle$ADP + $\angle$PDC = 180°
But $\angle$DAP + $\angle$ADP = 90°; therefore,
$\angle$DAP + $\angle$PDC = 90°.

It follows that $\angle$ADP = $\angle$PDC. Now extend AP a segment PE to equal itself AP = PE. Since the two right triangles DAP and DEP are congruent, point E is on DC. Now link and extend CE to meet the perpendicular line Px at D.

Link AD and draw other sides of the parallelogram ABCD.

Problem 4 of the Austrian Mathematical Olympiad 2003

In a parallelogram ABCD, let E be the midpoint of the side AB and F the midpoint of BC. Let P be the intersection point of the lines EC and FD. Show that the segments AP, BP, CP and DP divide the parallelogram into four triangles with areas in 1 : 2 : 3 : 4 ratio.

Solution

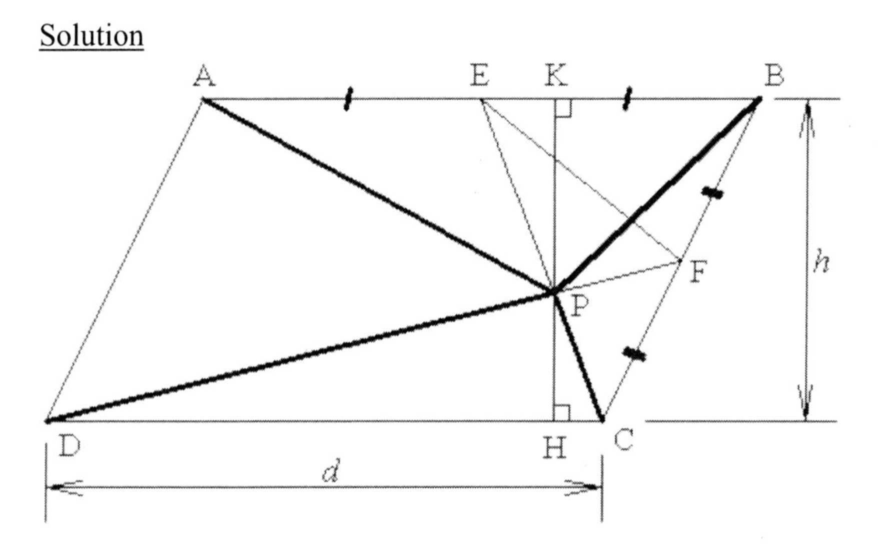

Let the horizontal side length of the parallelogram ABCD AB=CD = d, the vertical distance from AB to CD being h. Let (Ω) denote the area of shape Ω. It's easily seen that (DFC) = (EBC) = ¼ (ABCD). But the (DFC) and (EBC) share the common (PFC),

So now
(DPC) = (EBFP) = (EBP) + ½ (BPC) = (EBP) + ½ [(EBC) – (EBP)] (i)

Draw the altitudes from P to both AB and DC with their feet being K and H, respectively. Equation (i) can now be expressed as

$$\tfrac{1}{2} d \times PH = \tfrac{1}{4} d \times PK + \tfrac{1}{2} \left(\tfrac{1}{4} d \times h - \tfrac{1}{4} d \times PK \right) \quad \text{(ii)}$$

but PK + PH = h, and (ii) becomes

$4d \times PH = d \times PK + d \times h$ or $4(h - PK) = PK + h$

and $PK = 3h/5$, $PH = 2h/5$.

Therefore, $(DPC) = hd/5$ $(ABP) = 3hd/10$

Now $(BPC) = \frac{1}{4} d \times h - \frac{1}{4} d \times PK = hd/10$
$(ADP) = (ABCD) - (ABP) - (DPC) - (BPC) = 2hd/5$

or $(BPC) : (DPC) : (ABP) : (ADP) = 1 : 2 : 3 : 4$

Problem 4 of the Canadian Mathematical Olympiad 1976

Let AB be a diameter of a circle, C be any fixed point between A and B on this diameter, and Q be a variable point on the circumference of the circle. Let P be the point on the line determined by Q and C for which AC/CB = QC/CP. Describe, with proof, the locus of the point P.

Solution

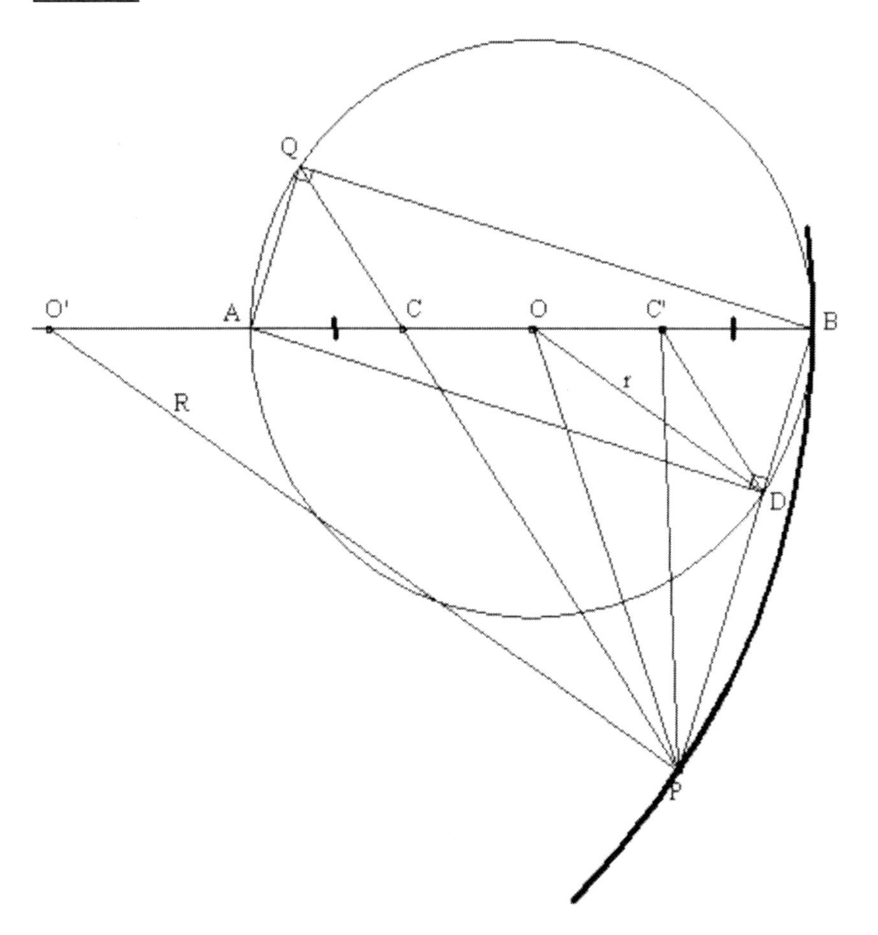

Let D be the intersection of the circle and BP. From AC/CB = QC/CP we have AC/QC = CB/CP, and triangles ACQ and BCP are similar since we also have

$\angle ACQ = \angle BCP$.

The similarity of the triangles gives us AC/CB = QC/CP = AQ/BP, and AQBD is a rectangle and AQ = BD, $\angle AQP = \angle BPQ$, $\angle QAB = \angle ABP$, AQ || BP and QB || AD.

Now pick the point C' as the image of point C across center O of the circle. We have AC/CB = BC'/CB = AQ/BP = BD/BP

Let r and O be the diameter and center of the circle, respectively. Link OD and from P draw a line to parallel with OD to meet AB extension at O'.

We have

OD/O'P = BD/BP = AC/CB or O'P = OD × CB/AC = r CB/AC and OB/O'B = OD/O'P = AC/CB, or O'B = O'P.

We conclude that the locus of the point P is a circle (only portion bold arc shown) with center at O' and radius R = O'P = r CB/AC.

Problem 4 of the Canadian Mathematical Olympiad 1978

The sides AD and BC of a convex quadrilateral ABCD are extended to meet at E. Let H and G be the midpoints of BD and AC, respectively. Find the ratio of the area of the triangle EHG to that of the quadrilateral ABCD.

Solution

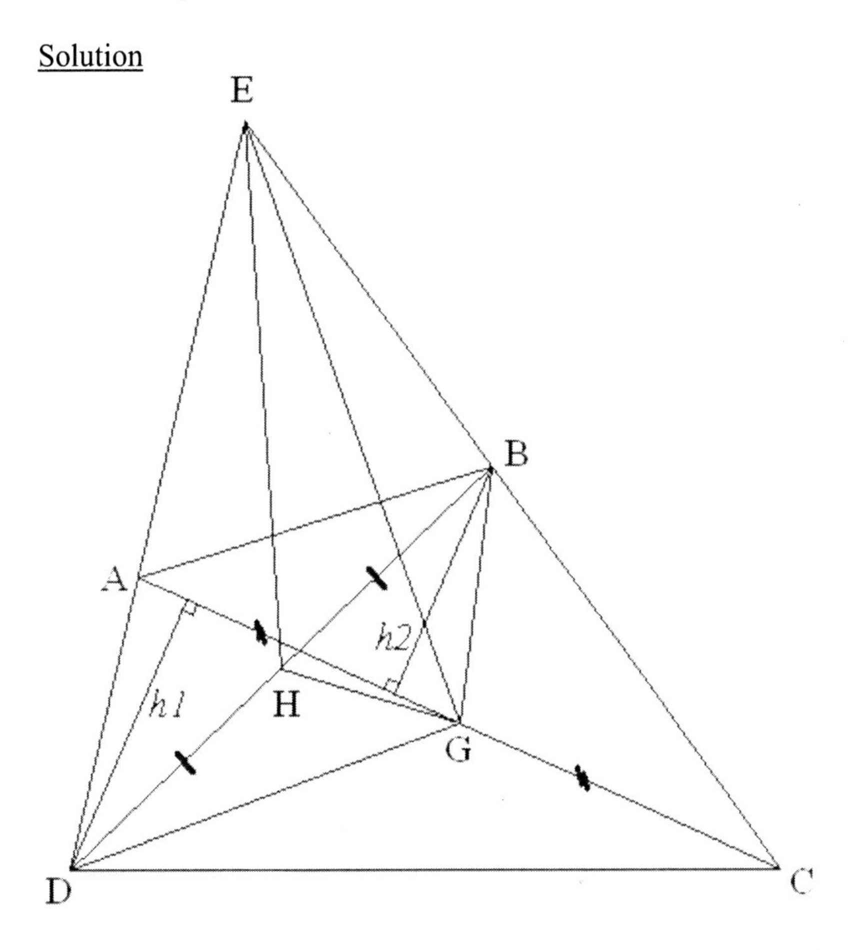

Let (Ω) denotes the area of shape Ω. Since H is the midpoint of BD,

(EHD) = (EHB) and (GHD) = (GHB) or
(EHG) + (EBG) = (EHD) + (GHD) (i)

Similarly, since G is the midpoint of AC,

(ABD) + (GBD) = (BGC) + (DGC)

or (ABD) + (GBD) = (ABGD) = ½ (ABCD)

But in quadrilateral EBGD, we have
(EHG) = ½ (ABCD) + (EAB) – (EBG) – [(EHD) + (GHD)]

Substituting (EHD) + (GHD) from (i), we have

(EHG) = ½ (ABCD) + (EAB) – (EBG) − (EHG) − (EBG) (ii)

But again since G is the midpoint of AC, the altitude from G to EB is half the altitude from A to EB, or (EAB) = 2(EGB).

Equation (ii) becomes

(EHG) = ½ (ABCD) + 2(EBG) – (EBG) − (EHG) − (EBG)

or
(EHG) = ½ (ABCD) − (EHG) or 2(EHG) = ½ (ABCD)

or
(EHG) / (ABCD) = ¼.

Problem 4 of the Ibero-American Mathematical Olympiad 1989

The incircle of the triangle ABC, is tangential to both sides AC and BC at M and N, respectively. The angle bisectors of the angles A and B intersect MN at points P and Q, respectively. Let O be the incenter of the triangle ABC. Prove that $MP \times OA = BC \times OQ$.

Solution

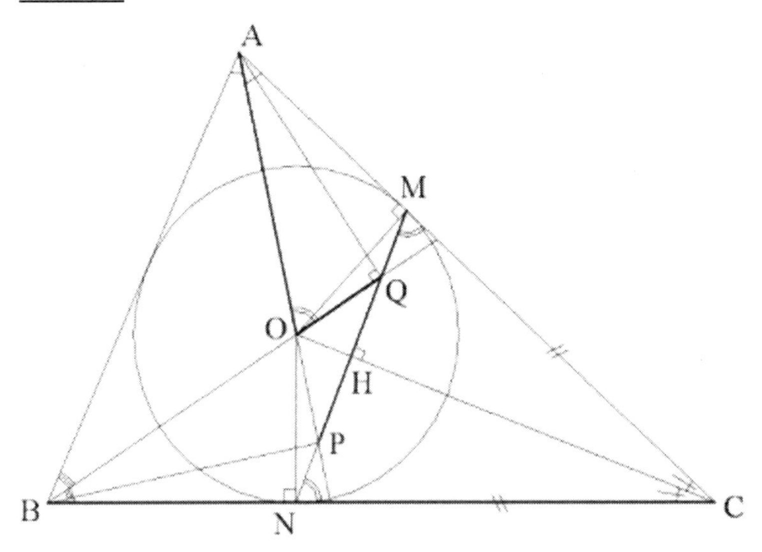

We have

$\angle AOQ = \angle ABO + \angle BAO = ½(180° - \angle C) = \angle HMC = \angle MOC$

and $\angle OMQ = \angle MCO$ (2 sides perpendicular to each other)

or $\angle AOQ + \angle AMQ = \angle HMC + 90° + \angle MCO = 180°$

Therefore, AMQO is cyclic and $\angle AQO = \angle AMO = 90°$ and triangles AQO and CHM and MHO are all similar.

Similarly, $\angle APB = 90°$.

These similarities give us $\frac{OA}{OQ} = \frac{CM}{MH} = \frac{CN}{MH} = \frac{OM}{OH}$ (i)

On the other hand because $\angle APB = 90°$, APNB is cyclic and $\angle OBN + \angle OPN = 180°$, or

$\angle OBN = \angle OPH$ or the two triangles OBN and OPH are also similar,

We have $$\frac{BN}{PH} = \frac{ON}{OH} = \frac{OM}{OH} \qquad \text{(ii)}$$

Combining (i) and (ii), we have

$$\frac{OA}{OQ} = \frac{CN}{MH} = \frac{BN}{PH} = \frac{CN + BN}{MH + PH} = \frac{BC}{MP}$$

or

$$MP \times OA = BC \times OQ$$

Problem 4 of the Ibero-American Mathematical Olympiad 1990

Let C1 be a circumference, AB one of its diameters, t its tangent in B, and M a point on C1 distinct of A and B. It is constructed a circumference C2 tangent to C1 on M, and to the line t.
(a) Find the point of tangency P to t and C2, and the locus of the centers of the circumferences C2 when M varies.
(b) Show that there exists a circle that is always orthogonal to Γ_2, regardless of the position of M. Note: Two circumferences are orthogonal to each other if they intersect and the respective tangents to the point of intersection are orthogonal.

Solution

(a) Let p and x0 be the radii, P and O the centers of the *C1* and the *C2* circles, respectively.

We have

$$y0^2 = O'H^2 = O'O^2 - HO^2 = (p + x0)^2 - (p - x0)^2 = 4px0$$

Thus the locus is a parabola whose focus is (p, 0) and whose directrix is x = -p. This type of parabola is used in auto headlight designs when light source is placed at the focus O (p,0) where light reflects off its surface and goes directly straight out as shown on the graph.

(b) We have $O'J^2 = y0^2 + (2p - x0)^2 = 4p^2 + x0^2$

From J draw a line through center O' of C2 and intercept C2 at K and Q as shown.

We also have $JK \times JQ = (O'J - x0) \times (O'J + x0) = O'J^2 - x0^2 = 4p^2$ or $(2p)^2 = JN^2 = JK \times JQ$ or JN is tangent to C2, and, therefore, C3 is always orthogonal to C2.

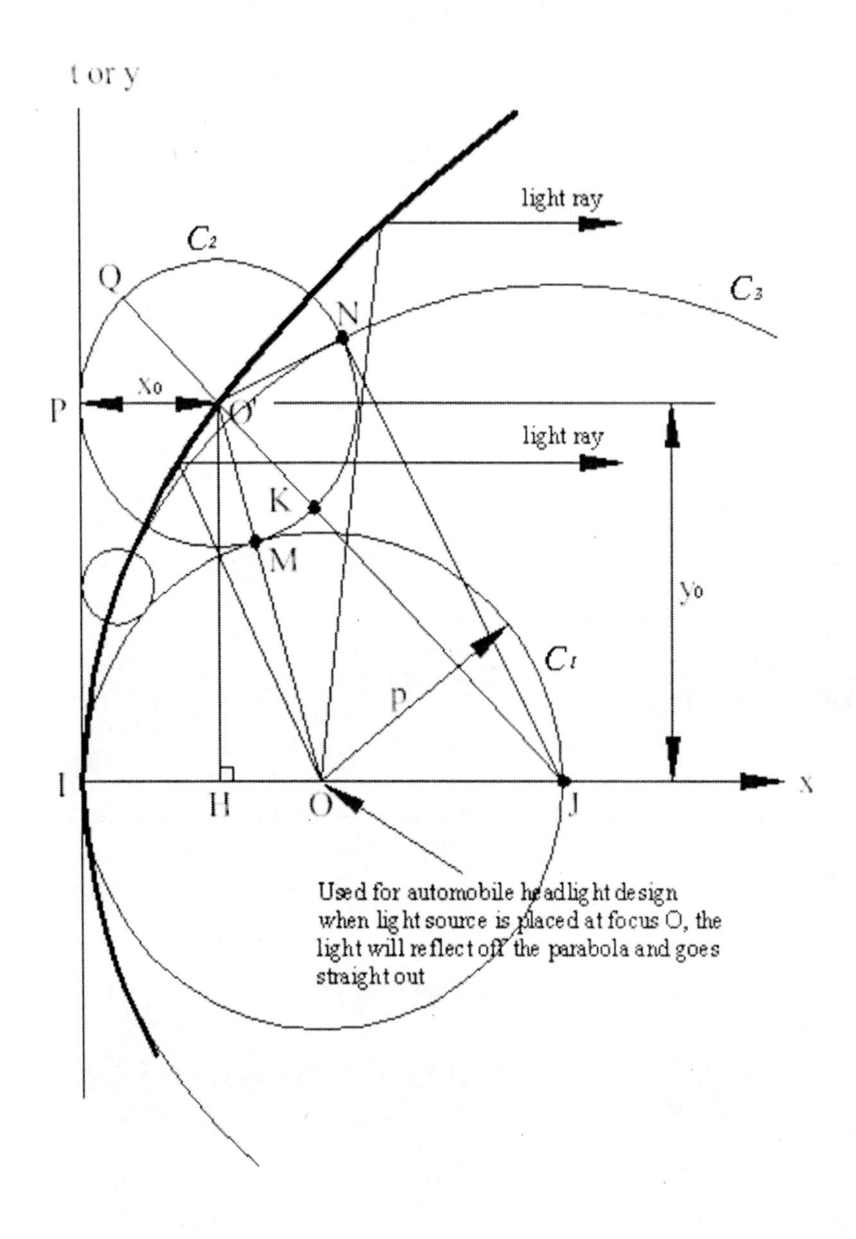
t or y
light ray
C2
Q
C3
N
x0
P
O'
light ray
K
M
y0
C1
p
I
H
O
J
x
Used for automobile headlight design when light source is placed at focus O, the light will reflect off the parabola and goes straight out

Problem 4 of the Ibero-American Mathematical Olympiad 1993

Let ABC an equilateral triangle and Γ its inscribed circle. If D and E are points in the sides AB and AC respectively, such that DE is tangent to Γ, show that

$$\frac{AD}{DB} + \frac{AE}{EC} = 1$$

Solution

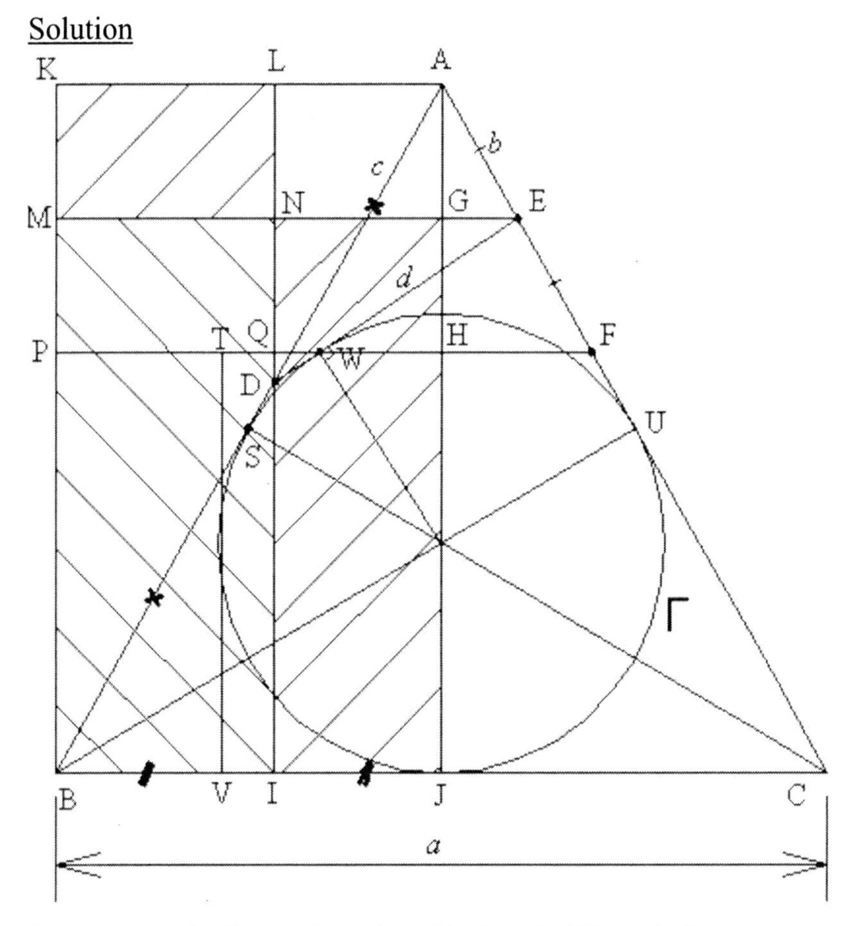

Let a, b, c and d be the lengths of BC, AE, AD and DE, respectively. Let J, U and S be the feet of A, B and C down to the sides of ABC. Let W be the tangential point of the circle and DE.

Pick F on AC so that EF = AE = b. From D draw perpendicular to and meet BC at I. From A, E and F draw horizontal lines to the left to meet the vertical lines from B and I at K, L, M, N, P and Q as shown.

We have AD/DB = IJ/BI and AE/EC = AG/GJ, so we have to prove IJ/BI +AG/GJ = 1, or

IJ × GJ + BI × AG = BI × GJ or to prove the areas of the two rectangles NGJI and KLNM equal the area of rectangle MNIB, or the area of rectangle PQIB = area of rectangle NGJI.

Pick point V on BJ so that BV = IJ, we now need to prove area of TQIV = area of NGHQ = area of LAGN.

We know IJ = ½ AD = c/2; therefore VI = a/2 – c and HJ = ½

$FC\sqrt{3} = \frac{1}{2}(a - 2b)\sqrt{3}$

and that $AG = \frac{1}{2} b\sqrt{3}$. Therefore, the area of

$TQIV = (a/2 - c) \times \frac{1}{2}(a-2b)\sqrt{3} = (\frac{1}{2} b\sqrt{3}) \times (c/2)$ is required to be proven.
or $a^2 - 2(b+c)a + 3bc = 0$ or

$$a = b+c + \sqrt{(b+c)^2 - 3bc}\,] = b+c + \sqrt{b^2 + c^2 - bc} \qquad \text{(i)}$$

But the cosine function for triangle ADE gives us $d^2 = b^2 + c^2 - bc$; equation (i) becomes

a = b + c + d which is obvious since d = DW + WE = DS + EU and b + c + d = AS + AU = a.

Problem 4 of the Ibero-American Mathematical Olympiad 1997

In an acute triangle ABC, let AE and BF be its altitudes, and H the orthocenter. The symmetric line of AE with respect to the angle bisector of angle A and the symmetric line of BF with respect to the angle bisector of angle B intersect each other on the point O. The lines AE and AO intersect again the circumscrite circumference to ABC on the points M and N respectively. Let P be the intersection of BC with HN; R the intersection of BC with OM; and S the intersection of HR with OP. Show that AHSO is a parallelogram.

Solution

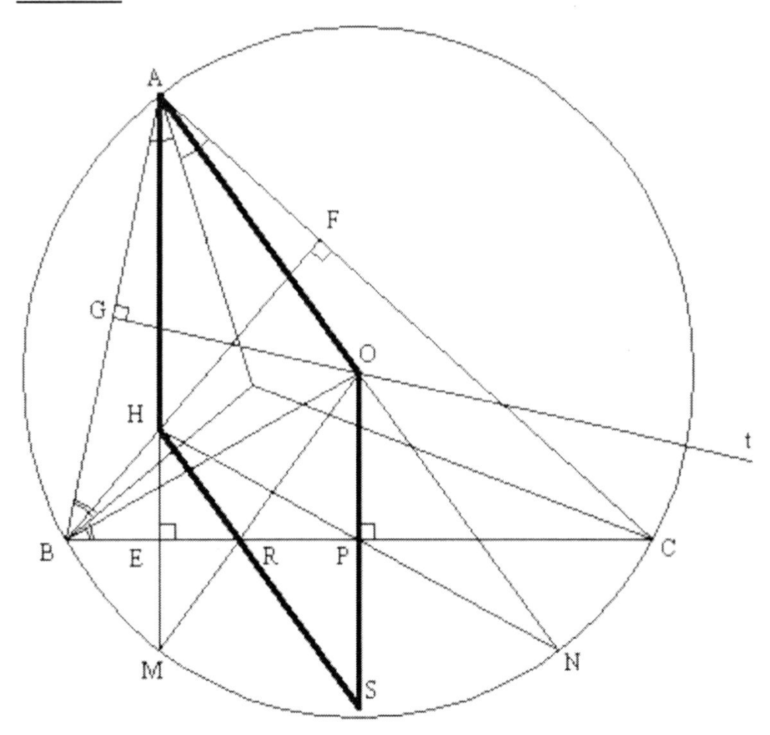

Let G be the midpoint of AB. Draw line *t* linking G and O. The problem gives us $\angle BAO = \angle EAF = \angle FBE$ (sides perpendicular) $= \angle ABO$ or $OA = OB$. So the center of the circumcircle is on line *t*.

The problem also gives us $\angle$BAM = $\angle$CAN and the arcs BM = CN or BC || MN and

$\angle$AMN = 90° and AN is, therefore, the diameter of the circumcircle. AN intercepts line *t* at O, and O is thus the center of the circumcircle.

Since H is the orthocenter, point M on the circle is, therefore, its image across BC.

We have HE = EM, and since BC || MN, HP = PN. P and O are midpoints of HN and AN, respectively, we have OP || AE which is one of the requirements for AHSO to be a parallelogram. The second requirement is for AO to parallel HS.

Since HM || OS and R is on symmetric segment BC, HS = OM = radius of the circle = OA.

Therefore, AO || HS which is the second requirement.

Problem 4 of the Ibero-American Mathematical Olympiad 2002

In a triangle ABC with all its sides of different length, D is on the side AC, such that BD is the angle bisector of $\angle ABC$. Let E and F, respectively, be the feet of the perpendicular drawn from A and C to the line BD and let M be the point on BC such that DM is perpendicular to BC. Show that $\angle EMD = \angle DMF$.

Solution

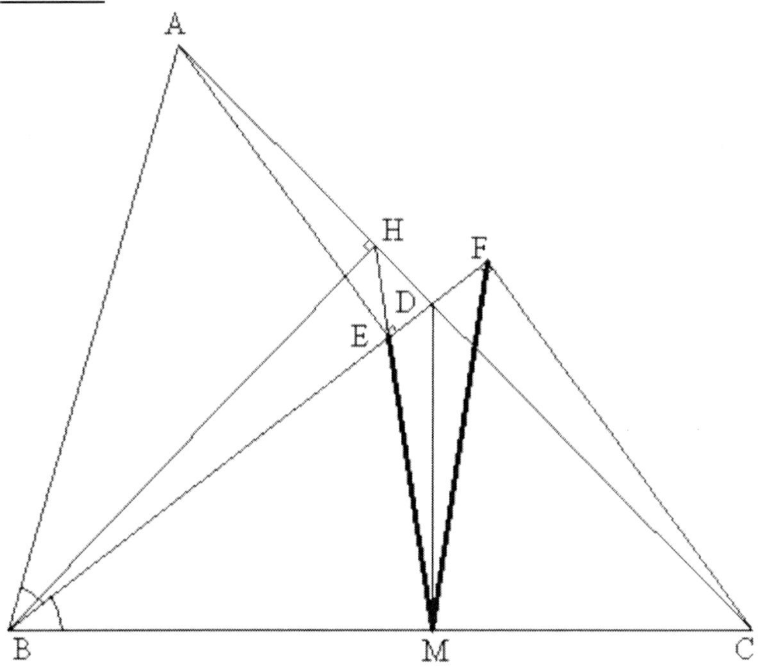

From B draw altitude to AC and meet it at H. We have the following cyclic quadrilaterals AHEB, BHDM and MDFC. Therefore,

$\angle BHE = \angle BAE = 90° - \frac{1}{2}\angle B = \angle BDM = \angle BHM$

$\angle BHE = \angle BHM$ therefore, H, E and M are collinear, and we have

$\angle EMD = \angle HMD = \angle HBD = \angle HBE = EAH = \angle EAD =$
$\angle DCF$ (AE, CF both $\perp$ EF) $= \angle DME$

Problem 4 of the International Mathematical Olympia 1960

Construct triangle *ABC;* given *ha; hb* (the altitudes from *A* and *B*) and *ma*, the median from vertex *A:*

Solution

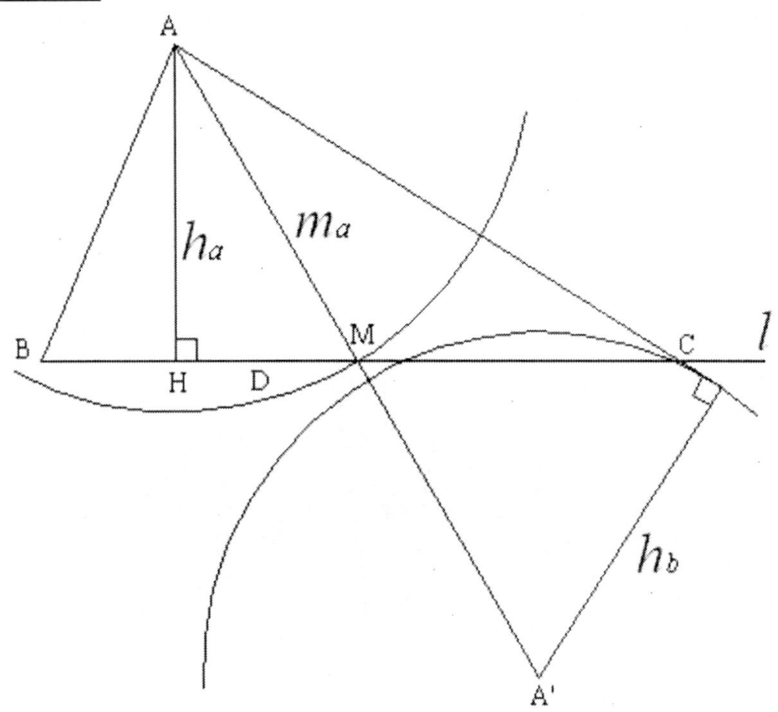

Draw line l. Both points B and C will be on this line. Pick an arbitrary point H on l. From H draw a segment HA perpendicular to l and with a length equal ha. Draw a circle with center A and radius ma to intercept line l at M. Extend AM and pick point A' at the extension so that MA' = MA. (A' is point symmetry of A across M).
Draw a circle *C* with center A' and radius hb. Then draw the tangential line from A to circle *C*. This tangential line will intercept line l at C. Point B is the image of C via point M.

Problem 5 of the Canadian Mathematical Olympiad 1969

Let ABC be a triangle with sides of lengths a, b and c. Let the bisector of the angle C cut AB in D. Prove that the length of CD is $2ab\cos(C/2) / (a + b)$

Solution

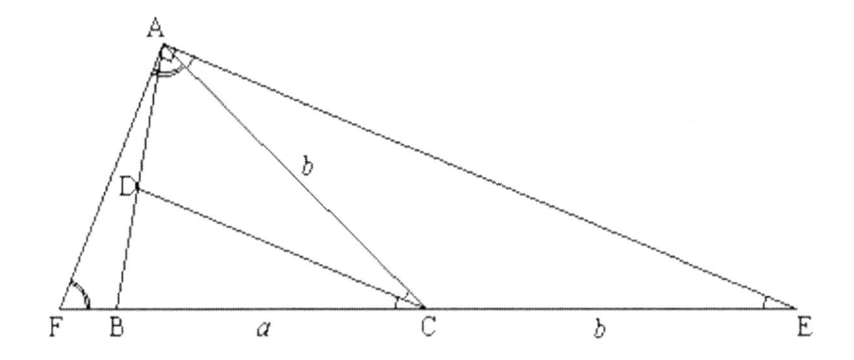

Extend BC to the right a length of CE = AC = b. From A draw the perpendicular to AE to meet the extension of CB to the left at F.

Since AC = CE, $\angle AEB = \frac{1}{2} \angle ACB = \angle DCB$ and $CD \parallel AE$ and we have $CD / AE = a / (a + b)$ or $CD = a\,AE / (a + b)$ (i)

We also have $\angle AFE = 90° - \angle AEF = 90° - \angle CAE = \angle FAC$ or CAF is isosceles with CA = CF = b and $\cos(C/2) = \cos\angle AEF = AE / (2b)$,

or $AE = 2b\cos(C/2)$, and (i) now becomes

$CD = 2ab\cos(C/2) / (a + b)$

Problem 5 of the Canadian Mathematical Olympiad 1970

A quadrilateral has one vertex on each side of a square of side-length 1. Show that the lengths a, b, c and d of the sides of the quadrilateral satisfy the inequalities $2 \le a^2 + b^2 + c^2 + d^2 \le 4$

Solution

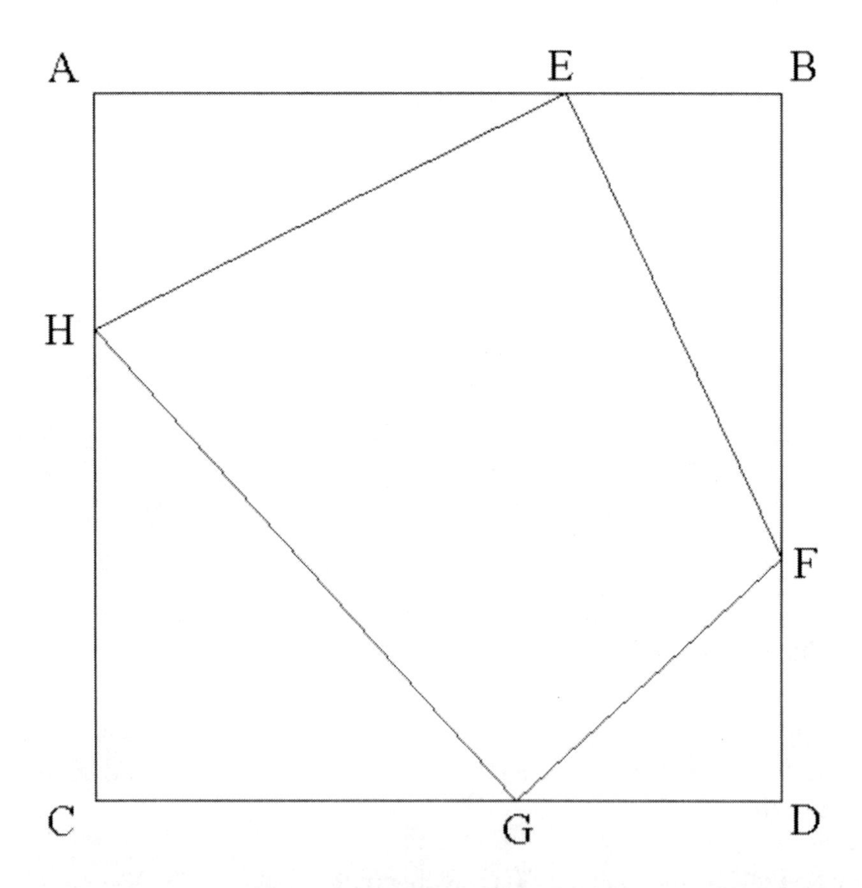

Let a = HE, b = EF, c = FG and d = GH. Based on problem 3 of Canadian Mathematical Olympiad year 1969, we have

$a^2 = HE^2 \ge (AH + AE)^2 /2$ and equality occurs when AH = AE = 0.5 which is half the length of the side of the square ABCD. So $a^2 + b^2 + c^2 + d^2 \ge 4 \times (.5 \times 2)^2 = 2$

It is also easily seen that

$a^2 + b^2 + c^2 + d^2 = AE^2 + EB^2 + BF^2 + FD^2 + DG^2 + GC^2 + CH^2 + HA^2$

And since $AE^2 + EB^2 \leq AE^2 + EB^2 + 2\ AE \times EB = (AE + EB)^2 = 1$ (equality occurs when either AE or EB = 0 or H or E coincides with the vertices of ABCD.

Similarly, $BF^2 + FD^2 \leq 1$, $DG^2 + GC^2 \leq 1$, $CH^2 + HA^2 \leq 1$ and

$a^2 + b^2 + c^2 + d^2 \leq 4$.

Problem 5 of the Canadian Mathematical Olympiad 1972

Prove that the equation $x^3 + 11^3 = y^3$ has no solution in positive integers x and y.

Solution

Rearrange the equation to $(x - y)^3 + 3x^2y - 3xy^2 = -11^3$ or

$(y - x) [(y - x)^2 + 3xy] = 11^3$

Since 11 is a prime integer, $y - x$ will take on the possible values of 1, 11, 11^2 or 11^3

If $y - x = 1$ then $3xy = 11^3 - 1 = 1330$ or $xy = 1330/3$ which is not an integer and therefore, this is not a possible scenario.

If $y - x = 11$ then $3xy = 0$ and either x or y must be 0 and not positive as required.

If $y - x = 11^2$ then $3xy = 11 - 11^2$ or $xy < 0$ and either x or y must be negative and not both being positive as required.

If $y - x = 11^3$ then $3xy = 1 - 11^3$ or $xy < 0$ which is the same as the previous case.

Problem 5 of the Canadian Mathematical Olympiad 1974

Given a circle with diameter AB and a point X on the circle different from A and B, let ta, tb and tx be the tangents to the circle at A, B and X respectively. Let Z be the point where line AX meets tb and Y the point where line BX meets ta. Show that the three lines YZ, tx and AB are either concurrent (i.e., all pass through the same point) or parallel.

Solution

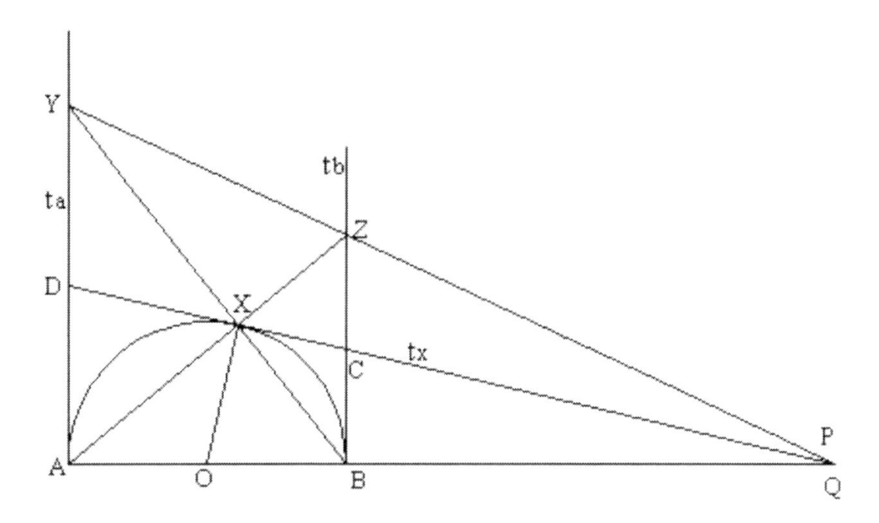

Let C and D be intersections of tx with BZ and AY, respectively. Also let YZ and DC to meet the extension of AB at P and Q, respectively.

Assume P and Q do not yet coincide.
Since BZ || AY, we have BZ / AY = BP / AP
but BZ / AY = BX / YX = CX / DX
DA and DX tangent the circle at A and X, respectively, we have DA = DX. The same applies to CB and CX and CB = CX, so

CX / DX = CB / DA = BQ / AQ = BP / AP or P coincides Q and the three lines YZ, tx and AB are concurrent. They will parallel one another when X is the midpoint of arc AB.

Problem 5 of the Canadian Mathematical Olympiad 1975

A, B, C, D are four "consecutive" points on the circumference of a circle and P, Q, R, S are points on the circumference which are respectively the midpoints of the arcs AB, BC, CD, DA. Prove that PR is perpendicular to QS.

Solution

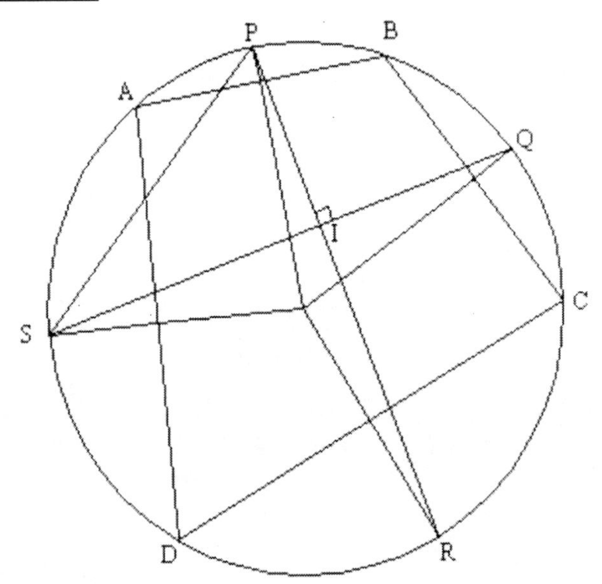

Let I be the intersection of PR and QS. We have $\angle PSQ$ subtending arcs PB and BQ; $\angle SPQ$ subtending arcs SD and DR. But since

arc PB = ½ arc AB, arc BQ = ½ arc BC, arc SD = ½ arc AD and arc DR = ½ arc DC

or $\angle PSQ$ and $\angle SPR$ combined to cut ½ the circle. Therefore,

$\angle PSQ + \angle SPR = 90°$, and $\angle SIP = 180° - \angle PSQ - \angle SPR = 90°$

or PR is perpendicular to QS.

Problem 5 of the Ibero-American Mathematical Olympiad 1992

Let Γ be a circle and let h and m be positive numbers such that there exists a trapezoid ABCD inscribed in Γ of height h and such that the sum of the bases AB+CD is m. Construct the trapezoid ABCD.

Solution

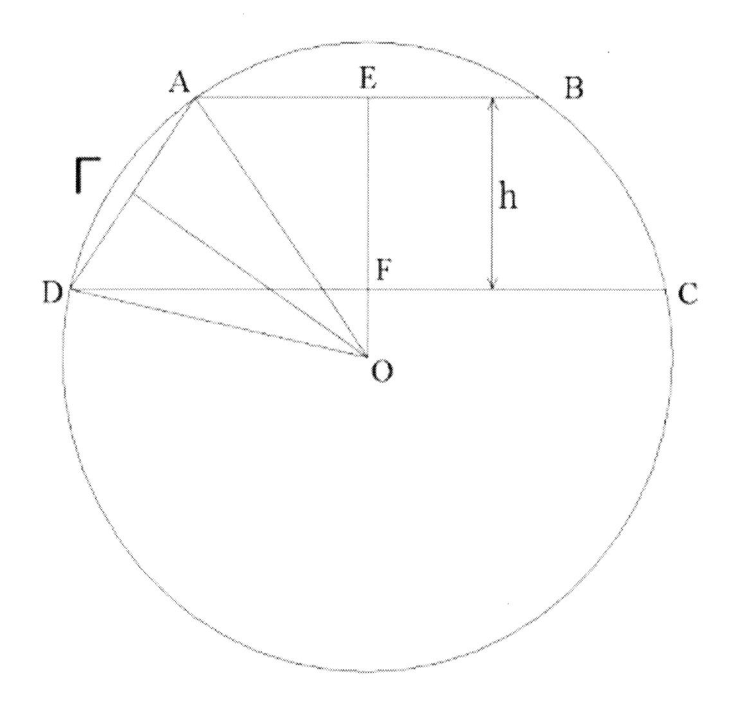

Let E and F be the midpoints of AB and DC, respectively and O be the center of circle Γ. Also let R be the radius of the circle. We have OA = OD = R and

$$AE^2 = R^2 - OE^2 = R^2 - (h + OF)^2 \qquad \text{(i)}$$

but $OF^2 = R^2 - DF^2 = R^2 - (m/2 - AE)^2$ or $OF = \sqrt{R^2 - (m/2 - AE)^2}$

Substituting OF to (i), we have

$$AE^2 = R^2 - [\, h + \sqrt{R^2 - (m/2 - AE)^2}\,]^2$$

we then need to solve the quadratic equation for AE

$$(m^2 + 4h^2)\,AE^2 - (m^3/2 + 2mh^2)\,AE + (h^2)^2 + (m^2)^2/16 - 4h^2R^2 + h^2m^2/2 = 0$$

This quadratic equation may not have a solution as we can see especially if R is too large compared to h and m.

Problem 5 of the Ibero-American Mathematical Olympiad 1999

An acute triangle 4ABC is inscribed in a circumference of center O. The highs of the triangle are AD; BE and CF. The line EF cut the circumference on P and Q.
(a) Show that OA is perpendicular to PQ.
(b) If M is the midpoint of BC, show that $AP^2 = 2AD \times OM$.

Solution

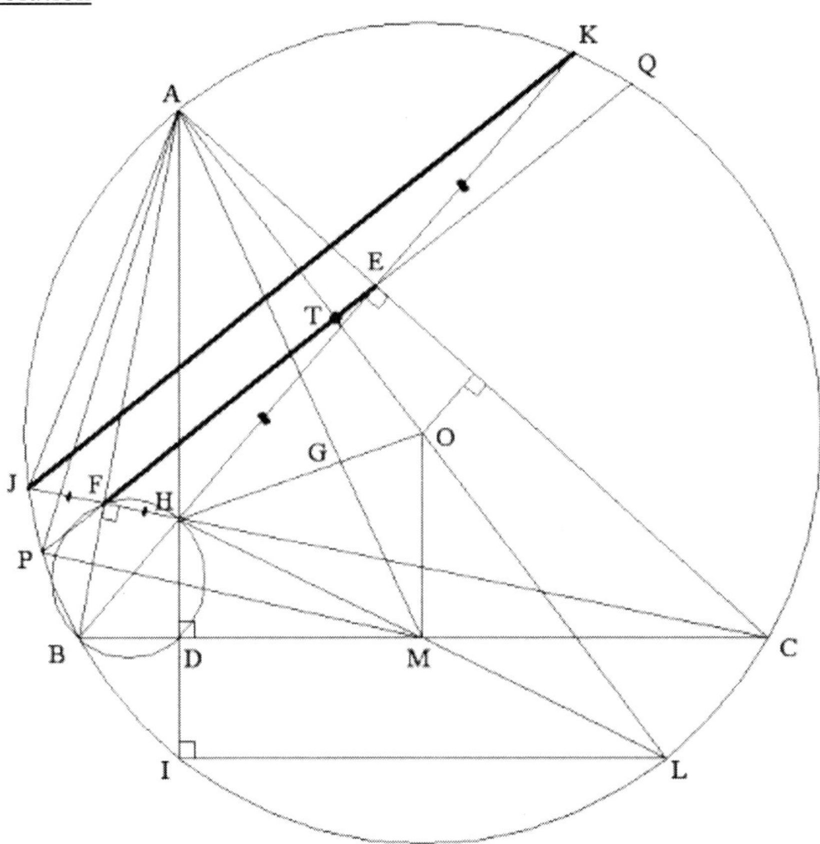

(a) Extend HE, HF and HD to meet the circle at K, J and I, respectively. Since H is the orthocenter of triangle ABC, the three points K, J and I are images of H across the three sides AC, AB and BC, respectively of the triangle ABC.
Therefore,

HE = EK and HF = FJ or FE || JK and arc KQ = arc JP and $\angle$PEB (subtends arcs KQ and PB) = $\angle$JKB (subtends arc JB) = $\angle$JCB = $\angle$BAI.

Extend AO to meet the circle at L. Since AL is the diameter of the circle, AI $\perp$ IL and IL || BC and BI = CL and $\angle$BAI = $\angle$CAL = $\angle$PEB.

$\angle$CAL = $\angle$PEB and AC $\perp$ BE, therefore, OA $\perp$ PQ.

(b) Let PQ intersect AO at T. Since OA $\perp$ PQ, we have $AP^2 = AT^2 + PT^2 = AT^2 + (PF + FT)^2 = AT^2 + PF^2 + FT^2 + 2PF \times FT = AT^2 + PF(PF + FT) + FT^2 + PF \times FT = AT^2 + PF \times PT + FT^2 + PF \times FT = AT^2 + PF(PT + FT) + FT^2 = AT^2 + PF \times FQ + FT^2 = AT^2 + AF \times FB + FT^2 = AF^2 + AF \times FB = AF \times (AF + FB) = AF \times AB$.

But FHDB is cyclic and we have AF × AB = AH × AD = 2 AD × OM since triangles AHG and MOG are similar and G is the centroid of triangle ABC and MG = ½ AG.

Problem 5 of the International Mathematical Olympiad 1959

An arbitrary point *M* is selected in the interior of the segment *AB:* The squares *AMCD* and *MBEF* are constructed on the same side of *AB;* with the segments *AM* and *MB* as their respective bases. The circles circumscribed about these squares, with centers *P* and *Q;* intersect at *M* and also at another point *N1:* Let *N* denote the point of intersection of the straight lines *AF* and *BC:*
(a) Prove that the points *N1* and *N* coincide.
(b) Prove that the straight lines *MN* pass through a fixed point *S* independent of the choice of *M:*
(c) Find the locus of the midpoints of the segments PQ as M varies between A and B:

Solution

(a) Two right triangles AFM and CBM are congruent because their sides are equal AM = CM and MF = MB. Therefore, $\angle$AFM = $\angle$CBM but since FM is perpendicular with BM, AF must perpendicular with CB or $\angle$ANB = 90 degree as always. So the locus of N is half the circle on top of diameter AB.

AC and FB are the two diagonal lines of the two squares AMCD and MBEF, respectively, and thus they are also diameters of the two circumcircles of those two squares, respective. Moreover, $\angle$ANC = 90 degree and $\angle$FNB = 90 degree, therefore, N is on both the circumcircles of those two squares or N1 coincides with N.

(b) As mentioned earlier, the locus of N is the top half circle with diameter AB. $\angle$ANM = $\angle$ACM = $\angle$MNB = $\angle$MFB = 45 degree, or $\angle$ANS = 45 degree and since N is on the circle, arc AS = ¼ the circumference of the circle with diameter AB and since A is fixed, S is also a fixed point independent of location of M.

(c) Let T be the midpoint of AB. Draw the three perpendicular lines PU, TR and QV to AB. We have:

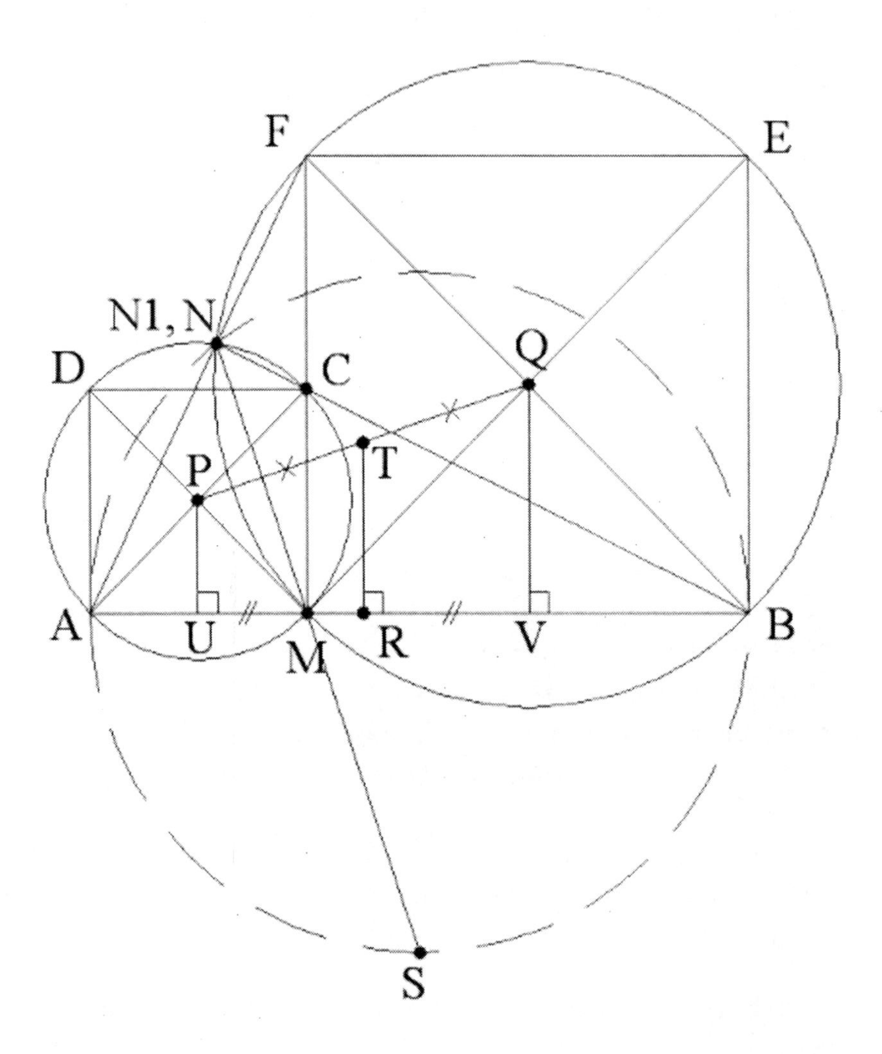

TR = ½(PU + QV) = ¼ (CM + EB) = ¼ AB

Therefore, when M moves between A and B the distance from the midpoint of PQ to AB is constant and is equal to ¼ AB. The locus is a straight segment parallel to AB, ¼ AB away from AB, and its length is equal to ½ AB the leftmost point of the locus is when M–>A; the rightmost is when M–>B. The segment that connects the midpoint of locus and midpoint of AB is perpendicular to AB.

Problem 6 of the Canadian Mathematical Olympiad 1971

Show that, for all integers n, $n^2 + 2n + 12$ is not a multiple of 121.

Solution

Assume $n^2 + 2n + 12$ is a multiple of 121. We have

$(n + 1)^2 + 11 = 121k$ where k is an integer, or

$(n + 1)^2 = 121k - 11 = 11\,(11k - 1)$ since 11 is a prime integer, for $(n + 1)^2 = 11\,(11k - 1)$ to occur we must have $11k - 1 = 11\,m^2$ where m is also an integer, or

$11\,(m^2 - k) = -1$, or

$m^2 - k = -1/11$ which is a fraction and is not possible. So the assumption that $n^2 + 2n + 12$ is a multiple of 121 is not possible.

Problem 6 of the Ibero-American Mathematical Olympiad 1987

Let ABCD be a plain convex quadrilateral. P, Q are points of AD and BC respectively such that $\frac{AP}{PD} = \frac{AB}{DC} = \frac{BQ}{QC}$.

Show that the angles that are formed by the lines PQ with AB and CD are equal.

Solution

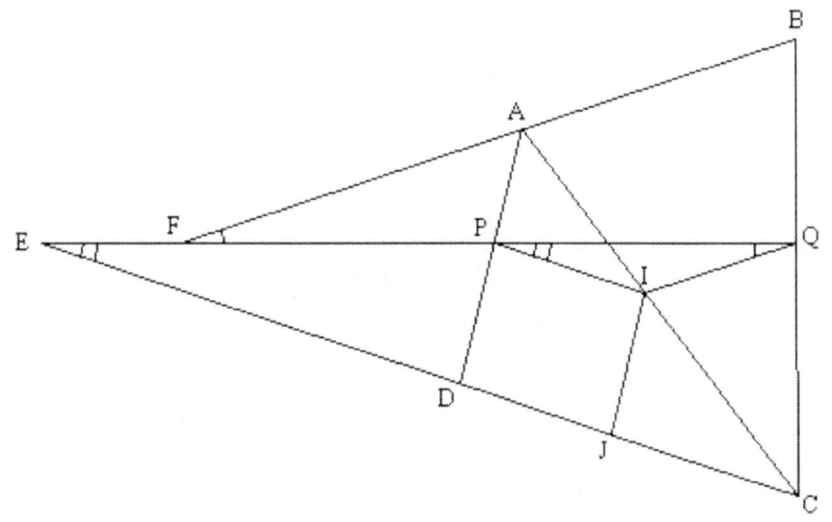

From P draw a line || to DC and intercept AC at I. Link IQ. We have IQ || AB. We then have $\angle QEC = \angle QPI$ and $\angle QFB = \angle PQI$
To prove that the angles that are formed by the lines PQ with AB and CD are equal,
we then need to prove $\angle QPI = \angle PQI$ or $IP = IQ$.

From I draw a line || to AD and intercept DC at J. We have
$\frac{IQ}{AB} = \frac{IC}{AC} = \frac{JC}{DC}$ or $\frac{AB}{DC} = \frac{IQ}{JC}$

We also have $\frac{AB}{DC} = \frac{AP}{PD} = \frac{IP}{JC}$; therefore, $IP = IQ$.

Problem 6 of the USA Mathematical Olympiad 1999

Let ABCD be an isosceles trapezoid with $AB \parallel CD$. The inscribed circle w of triangle BCD meets CD at E. Let F be a point on the (internal) angle bisector of $\angle DAC$ such that $EF \perp CD$. Let the circumscribed circle of triangle ACF meet line CD at C and G. Prove that the triangle AFG is isosceles.

Solution

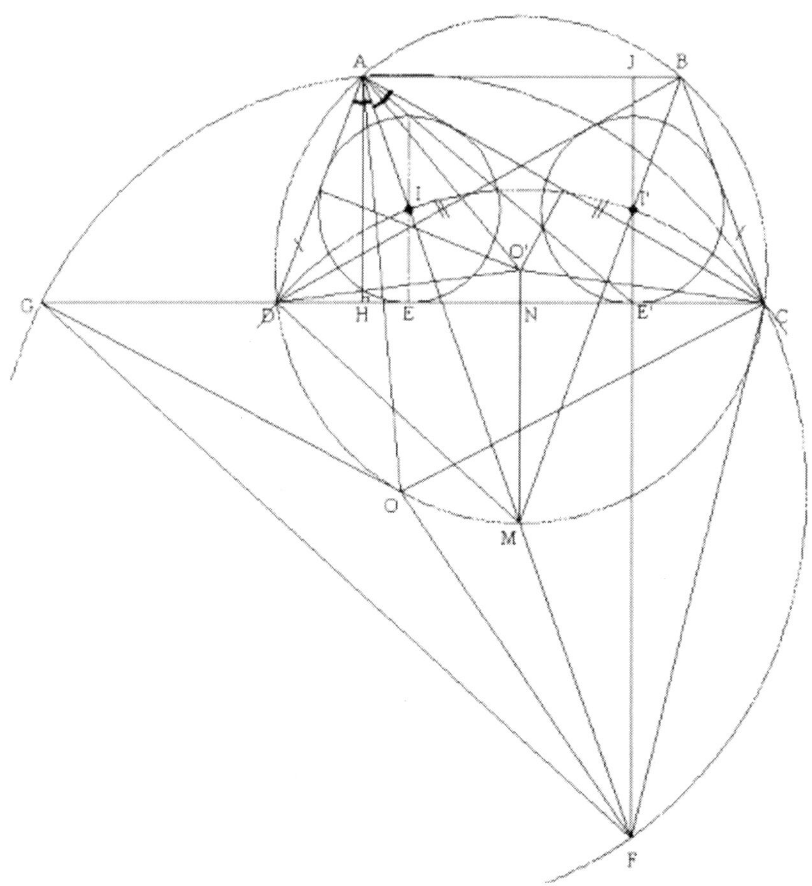

Let w be the circumcircle of triangle ACF.
Draw the incircle of triangle ADC with center at I'; this circle is symmetrical of the incircle of triangle BDC via the axis passing through centers of AB and DC.

Draw the circumcircle w1 of triangle ADC to intercept AF at M.

Since AF is the bisector of $\angle DAC$, we have MD = MI' = MI, and M is the center of circle w2 as shown.

From M draw line perpendicular to DC and meets it at N. Since N is the midpoint of EE', MI' = MF and therefore, F is on circle w2.

For circle w1, we have: $AP \times PM = DP \times PC$ (i)

For circle w, we have: $AP \times PF = GP \times PC$ (ii)

From (i) and (ii) $PM / PF = DP / GP$ (iii)

or $MD \parallel GF$ and
$MD / GF = PM / PF$

or $GF = MD \times PF / PM$ (iv)

For circle w2, we have: $IP \times PF = DP \times PC$ (v)

From (v) and (ii), we have: $IP / AP = DP / GP$ (vi)

From (vi) and (iii), we have: $IP / AP = PM / PF$ (vii)

From (vii)
$IP / AP = PM / PF = (IP+PM) / (AP+PF) = MI / AF$ (viii)

From (iv) and (viii) $GF = MD \times AF / MI = AF$

Therefore, triangle AFG is isosceles.

Problem 6 of the Australian Mathematical Olympiad 2008

Let $A_1A_2A_3$ and $B_1B_2B_3$ be triangles. If $p = A_1A_2+A_2A_3+A_3A_1+B_1B_2+B_2B_3+B_3B_1$ and $q = A_1B_1+A_1B_2+A_1B_3+A_2B_1+A_2B_2+A_2B_3+A_3B_1+A_3B_2+A_3B_3$, prove that $3p \leq 4q$.

Solution

We know that in a triangle the sum of the two sides is greater than the third side. And in the case of a degenerate triangle, the sum is equal to length of the third side. Therefore, we have the following inequalities

$A_1B_1 + A_2B_1 \geq A_1A_2$

$A_1B_2 + A_2B_2 \geq A_1A_2$

$A_1B_3 + A_2B_3 \geq A_1A_2$

$A_2B_1 + A_3B_1 \geq A_2A_3$

$A_2B_2 + A_3B_2 \geq A_2A_3$

$A_2B_3 + A_3B_3 \geq A_2A_3$

$A_1B_1 + A_3B_1 \geq A_1A_3$

$A_1B_2 + A_3B_2 \geq A_1A_3$

$A_1B_3 + A_3B_3 \geq A_1A_3$

$B_1A_1 + B_2A_1 \geq B_1B_2$

$B_1A_2 + B_2A_2 \geq B_1B_2$

$B_1A_3 + B_2A_3 \geq B_1B_2$

$B_2A_1 + B_3A_1 \geq B_2B_3$

$B_2A_2 + B_3A_2 \geq B_2B_3$

$B_2A_3 + B_3A_3 \geq B_2B_3$

$B_1A_1 + B_3A_1 \geq B_1B_3$

$B_1A_2 + B_3A_2 \geq B_1B_3$

$B_1A_3 + B_3A_3 \geq B_1B_3$

Now adding all these inequalities, we have $3p \leq 4q$.

Problem 7 of the Australian Mathematical Olympiad 2009

Let I be the incenter of a triangle ABC in which $AC \neq BC$. Let Γ be the circle passing through A, I and B. Suppose Γ intersects the line AC at A and X and intersects the line BC at B and Y. Show that $AX = BY$.

Solution

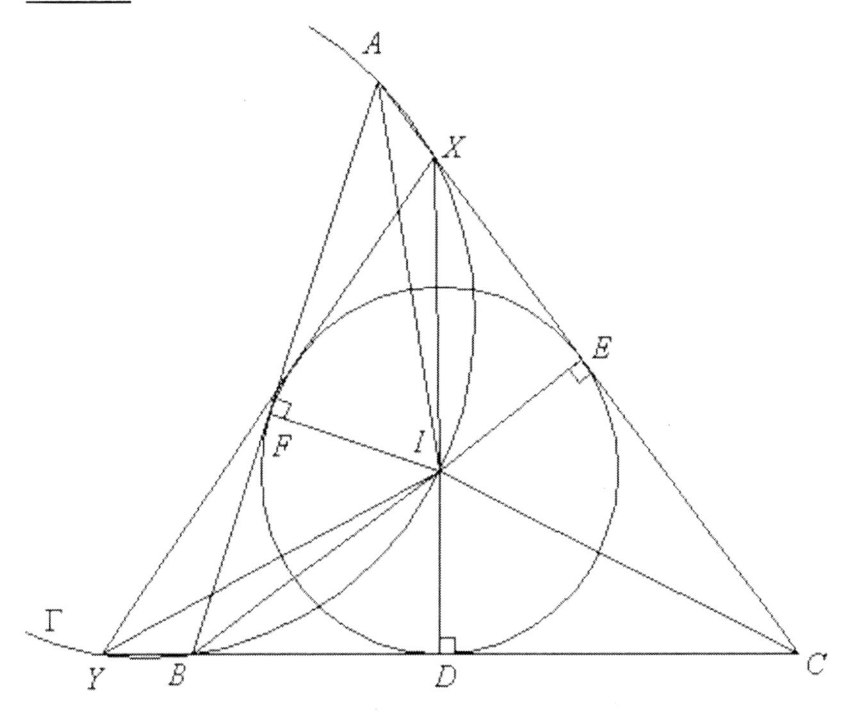

Since BI is the bisector of ∠ABC, we have ∠ABI = ∠CBI, but ∠CBI = ∠CYI + ∠YIB subtends arcs YB + BI = arc YI. Meanwhile, ∠ABI subtends arc AI. Therefore,

AI = YI and ∠ABI = ∠YXI (i)

We also have ∠AXY = ∠ABY (subtend larger arc AY) or ∠ABC = ∠YXC or ∠ABI + ∠CBI = ∠YXI + CXI.

Combining with (i), we have $\angle CBI = \angle CXI$ and the two triangles BDI and XEI being congruent, or $XE = BD$.

We also have $EC = DC$ and thus $CX = CB$.
From point C outside circle Γ, we have $CX \times CA = CB \times CY$ or $CA = CY$.

Hence, $AX = CA - CX = CY - CB = BY$

Problem 7 of Belarusian Mathematical Olympiad 2000

On the side AB of a triangle ABC with BC < AC < AB, points B1 and C2 are marked so that AC2 = AC and BB1 = BC. Points B2 on side AC and C1 on the extension of CB are marked so that CB2 = CB and CC1 = CA. Prove that the lines C1C2 and B1B2 are parallel.

Solution

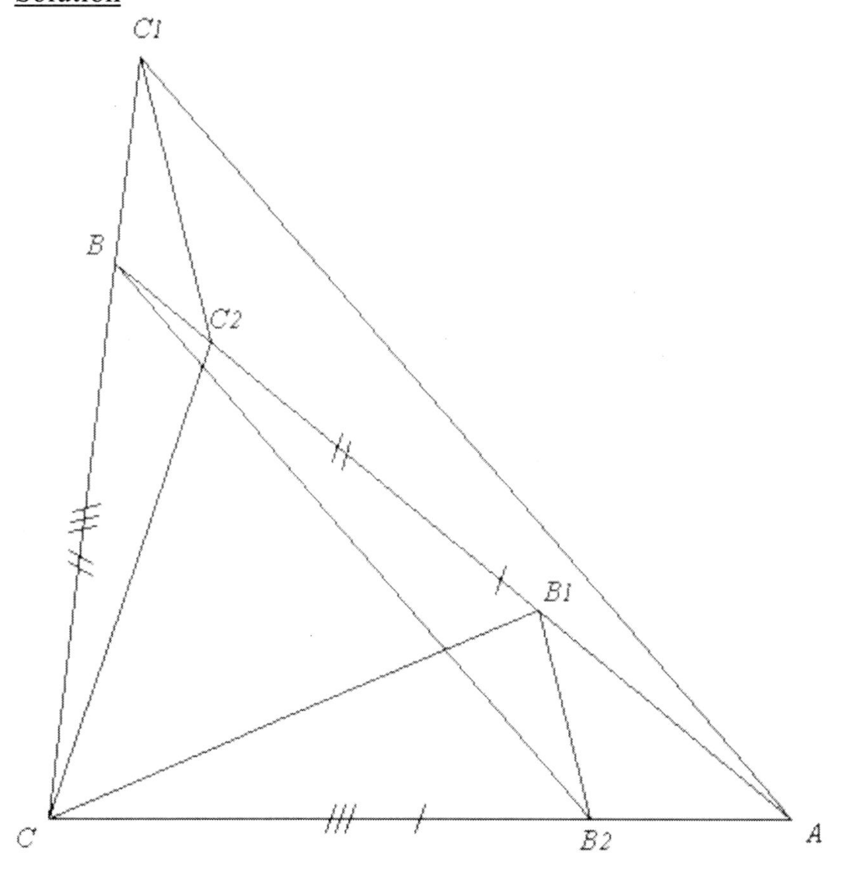

Observe that BB2 and C1A are parallel as given by the problem which makes $\angle$B2BB1 = $\angle$C1AC2. All we need to do now is to prove that the two triangles B2BB1 and C1AC2 are similar which makes $\angle$C2C1A = $\angle$B1B2B and C1C2 || B1B2.

Given $\angle$B2BB1 = $\angle$C1AC2 as mentioned, we only need to prove
$$\frac{BB2}{AC1} = \frac{BB1}{AC2}$$
But since BB2 || AC1, we have $\frac{BB2}{AC1} = \frac{CB}{CC1}$

The problem also gives CB = BB1 and CC1 = AC = AC2;

therefore, $\frac{CB}{CC1} = \frac{BB1}{AC2}$

or $\frac{BB2}{AC1} = \frac{BB1}{AC2}$ and the proof is done.

Problem 7 of Belarusian Mathematical Olympiad 2004

Let be given two similar triangles such that the altitudes of the first triangle are equal to the sides of the other. Find the largest possible value of the similarity ratio of the triangles.

Solution

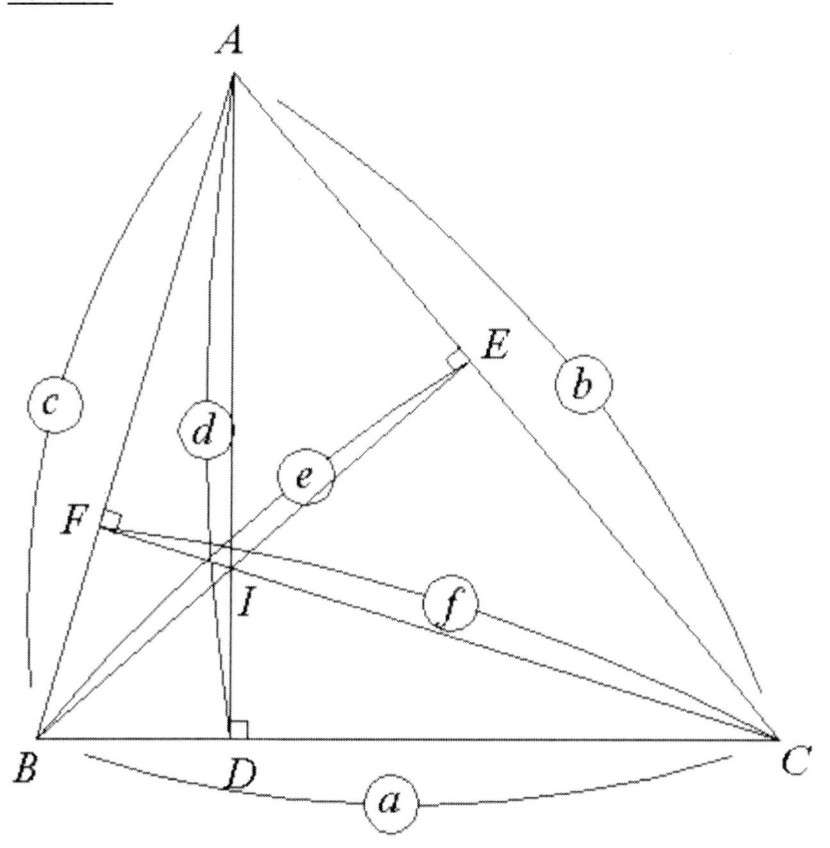

Let the first triangle be ABC and the feet from A, B and C to the opposite sides be D, E and F, respectively. Now let BC = a, AC = b, AB = c and AD = d, BE = e and CF = f.

Assume $a \geq b \geq c$. Because twice the area of triangle ABC = ad = be = cf, our assumption makes $f \geq e \geq d$.

To find the largest possible value of the similarity ratio of the triangles we need to find the largest possible ratio $\frac{f}{a}$ or largest possible $\cos\angle FCB$.

The similarity of the triangles ADB and CFB gives us

$$\frac{d}{c} = \frac{f}{a} \qquad \text{(i)}$$

and similarity of triangles AEB and AFC,

$$\frac{e}{c} = \frac{f}{b} \qquad \text{(ii)}$$

And because the altitudes of the first triangle are equal to the sides of the second, we also have

$$\frac{f}{a} = \frac{e}{b} \qquad \text{(iii)}$$

From (ii), $e = \frac{cf}{b}$; substituting it into (iii), we then have $b^2 = ac$

Now the law of cosines gives us $b^2 = a^2 + c^2 - 2ac \times \cos\angle ABC$, or

$ac = a^2 + c^2 - 2ac \times \cos\angle ABC$, or

$$\cos\angle ABC = \frac{a^2 + c^2 - ac}{2ac} = \frac{a^2 + c^2}{2ac} - \frac{1}{2}$$

But $\angle ABC + \angle FCB = 90°$; therefore,

$$\cos\angle FCB = \sin\angle ABC = \sqrt{1 - \cos^2\angle ABC}$$

Hence, $\cos\angle FCB$ is largest when $\cos^2\angle ABC$ is smallest or when $\frac{a^2 + c^2}{2ac} - \frac{1}{2}$ is smallest, or when $\frac{a^2 + c^2}{2ac}$ is smallest which happens when $a = c$ (per AM-GM inequality) which makes $\frac{a^2 + c^2}{2ac} = 1$.

Thus, the largest possible $\cos\angle FCB = \sqrt{1 - \frac{1}{4}} = \frac{1}{2}\sqrt{3}$ when

$a = b = c$ and the triangle ABC and its similar triangle are both equilateral.

Further observation

It depends on how one defines the similarity ratio; the similarity ratio could be the ratio of the side of the larger triangle to the corresponding side of the smaller one. In such a case, the similarity ratio is the inverse of the above result which is $2\sqrt{3}/3$. This ratio is the largest and could be the solution required.

Problem 7 of the Canadian Mathematical Olympiad 1969

Show that there are no integers a, b, c for which $a^2 + b^2 - 8c = 6$.

Solution

Adding 2ab to both sides, we have $(a + b)^2 = 2(ab + 4c + 3)$ or

$ab + 4c + 3 = 2d^2$ where d is an integer. Since $2d^2$ is even, the product ab must be an odd number and both **a and b must be odd numbers**.

Now let $a = 2m + 1$ and $b = 2n + 1$ where m and n are integers. Substituting them into the original equation, we have

$(2m + 1)^2 + (2n + 1)^2 = 2(4c + 3)$, or

$4m^2 + 4m + 4n^2 + 4n + 2 = 2(4c + 3)$, or

$m^2 + m + n^2 + n = 2c + 1$, or
$m(m + 1) + n(n + 1) = 2c + 1$ (i)

Now note that the product of two consecutive numbers is always an even number since one of them is an even number.

Therefore, the sum on the left of (i) is an even number whereas the one on the right is an odd number. So the original requirements of both a and b being odd numbers are also not possible.

Therefore, we can not find integers for a, b and c to satisfy the problem.

Problem X of Russia Mathematical Contest

Given triangle ABC, construct three pair-wise orthogonal circles that each pass through a pair of vertices.

Solution

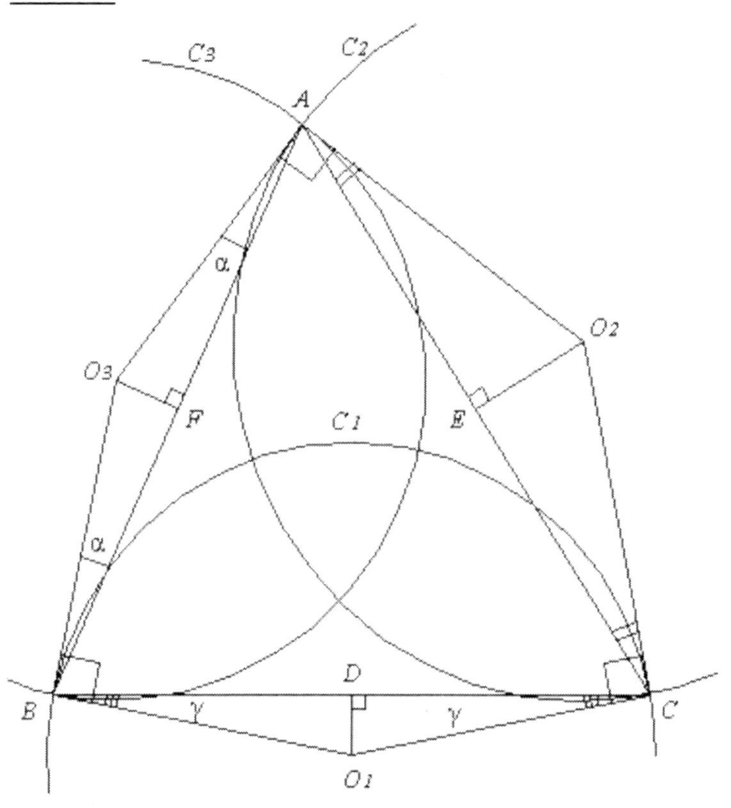

This problem is equivalent to finding three points on perpendicular bisectors such that three angles at the vertices through the three points are 90°.

Let's name these three points O1, O2 and O3, and the angle $\alpha = \angle O3AB = \angle O3BA$, $\beta = \angle O2AC = \angle O2CA$ and $\gamma = O1BC = \angle O1CB$.

We have $\angle O3AO2 + \angle O2CO1 + \angle O1BO3 = 90° \times 3 = 270°$ or

$2(\alpha+\beta+\gamma)+\angle BAC+\angle ACB+\angle CBA=270°$ or

$2(\alpha+\beta+\gamma)+180°=270°$ or $\alpha+\beta+\gamma=45°$

But $\alpha+\beta+\angle BAC=\angle O3AO2=90°$ or
$\angle BAC-45°=\gamma$

The other two angles α and β can be found by following the same procedure.

Further observation

This procedure only uses the graph with every angle of the triangle >45°, other configurations can be solved with similar approach.

Problem 2 of the Ibero-American Mathematical Olympiad 1991

Two perpendicular lines divide a square in four parts, three of them has area equal to 1. Show that the area of the full square is four.

Solution

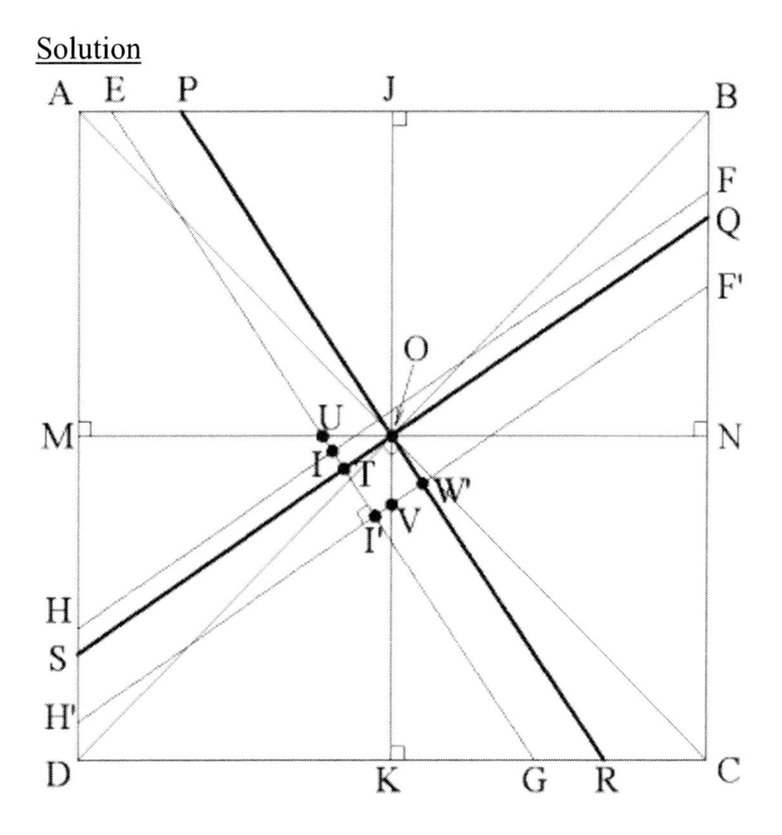

Let the square be ABCD and the two perpendicular lines dividing it into four parts intercept each other at I and let them divide the square into AEIH, DHIG, CGIF and BFIE whereas the areas of AEIH, DHIG, CGIF are all equal to 1.

Divide the square into four equal squared quadrants AJOM, MOKD, JBNO and ONCK. We only prove the problem with the location of I in a single quadrant MOKD. Note that there are four possible distinct locations for I to cover all the scenarios of the problem:

(1) I is located above line SQ or on MO (I on line HF)

(2) I is located on segment SO (I coincides with T)
(3) I is located below line SQ (I on line H'F')

Case (1) I is located above line SQ or on MO (I on line HF)

From O draw a line parallel to EG to intercept AB and DC at P and R, respectively and draw another line parallel to HF to intercept AD and BC at S and Q, respectively.

In two quadrilaterals APRD and CRPB, $\angle APR = \angle CRP$ and mid-segments OM = ON; therefore, the two quadrilaterals APRD and CRPB are congruent and their areas are equal.

The two quadrilaterals DSQC and BQSA are congruent and their areas are equal.

These two perpendicular lines PR and SQ divide the square into four equal areas since AP= CR=DS=BQ. Now let SQ intercept EG at T, MN intercept EG at U and let (Ω) denotes the area of Ω, we have

$$(AEIH) = (APOS) - (EPOU) - (UOT) - (HITS) \qquad \text{(i)}$$
$$(HIGD) = (SORD) - (UORG) + (UOT) + (HITS) \qquad \text{(ii)}$$

Since (AEIH) = (HIGD), (APOS) = (SORD) and (EPOU) = (UORG), subtract (i) from (ii) we have (UOT) + (HITS) = 0. This is impossible; therefore, this case is not possible.

Case (2) I is located on SQ (I coincides T and H coincides S, F coincides Q)

$$(AETS) = (APOS) - (EPOU) - (UOT) \qquad \text{(iii)}$$
$$(STGD) = (SORD) - (UORG) + (UOT) \qquad \text{(iv)}$$

Subtract (iii) from (iv) we have (UOT) = 0 this is also impossible; therefore, this case is also not possible.

Case (3) I is located below line SQ (I on line H'F')

Let V and W be the intersections of H'F' and OK, and H'F' and PR, respectively. We have

(H'I'GD) = (SORD) – (SOVH') – (VOW) – (GI'WR) (v)
(CGI'F') = (CROQ) – (VOQF') + (VOW) + (GI'WR) (vi)

Since (H'I'GD) = (CGI'F'), (SORD) = (CROQ) and (SOVH') = (VOQF'), subtracting (v) from (vi) yields (VOW) + (GI'WR) = 0 which is also impossible; therefore, this case is also not possible.

Therefore, the only prevailing scenario is for I to coincide with O and the four parts to be equal as proven earlier and the total area is 3 + 1 = 4.

Further observation

The problem below is derived from the above problem:
Two perpendicular lines divide a square into four parts, three of them have equal areas. Prove that all four parts have equal areas.

Problem 1 of Austrian Mathematical Olympiad 2004

Determine all integers a and b such that $(a^3 + b)(a + b^3) = (a + b)^4$

Solution

Expanding the equation and canceling the same terms, we have

$(b^2 - 4)a^2 - 6ba - 4b^2 + 1 = 0$

Solving for a, we have

$$a_1 = \frac{2b^2 + 3b - 2}{b^2 - 4} \quad \text{and} \quad a_2 = \frac{-2b^2 + 3b + 2}{b^2 - 4}$$

When $a_1 = \dfrac{2b^2 + 3b - 2}{b^2 - 4} = \dfrac{2b - 1}{b - 2}\ (b \neq -2) = 2 + \dfrac{3}{b - 2}\ (b \neq 2)$

which is an integer when b = 1, b = 3 and b = 5

When $a_2 = \dfrac{-2b^2 + 3b + 2}{b^2 - 4} = -\dfrac{2b + 1}{b + 2}\ (b \neq 2) = -2 + \dfrac{3}{b + 2}\ (b \neq -2)$

which is an integer when b = -1, b = -3 and b = -5

Answers: (a, b) = (-3, -5), (-5, -3), (-1, -1), (1, 1), (5, 3) and (3, 5).

Problem 2 of Austrian Mathematical Olympiad 2000

The trapezoid ABCD (ABCD labeld counterclockwise, AB ≠ CD) is inscribed into a circle k. On the arc AB two points P and Q (P ≠ Q) are chosen (with APQB labeld in counterclockwise order). Let X be the intersection of the lines CP and AQ and Y be the intersection of the lines BP and DQ.
Show that P, Q, X and Y lie on a circle.

Solution

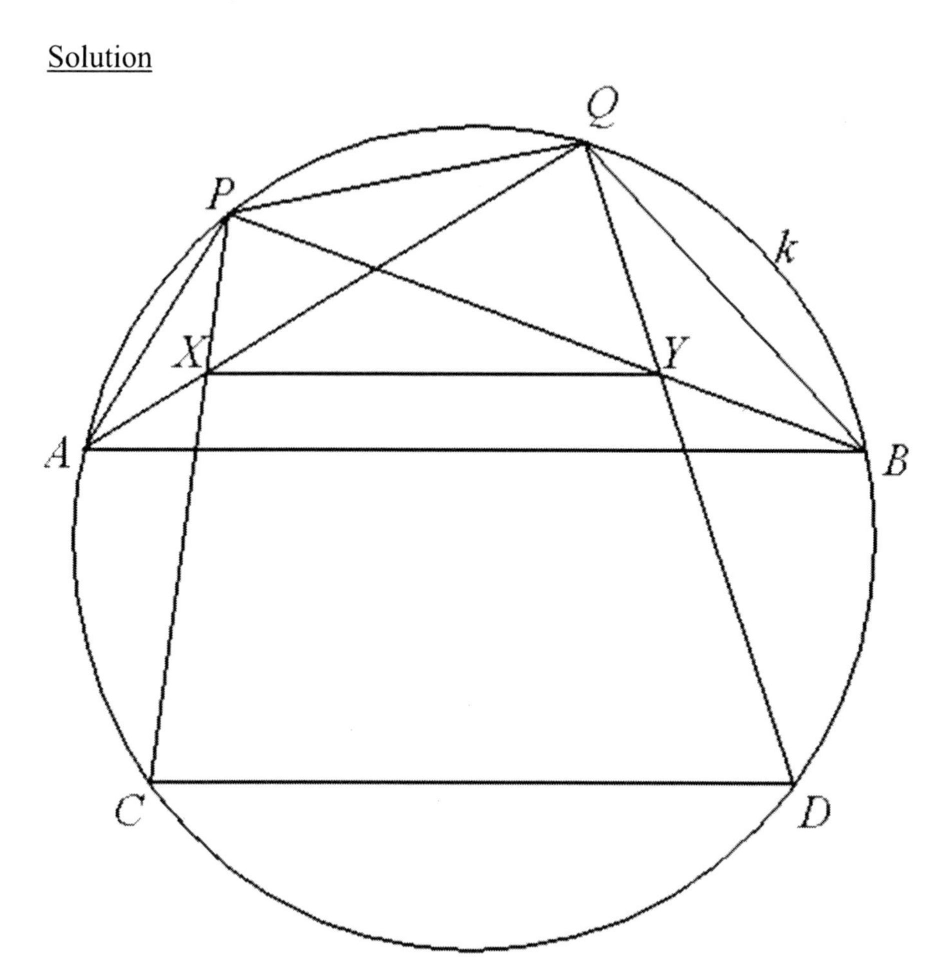

Since AB || CD, arc AC = arc BD, or arc AD = arc BC. ∠AQD subtending arc AD = ∠BPC subtending arc BC.

Therefore, PQYX are cyclic, and P, Q, X and Y lie on a circle.

Problem 2 of the Austrian Mathematical Olympiad 2005

A semicircle h with diameter AB and center M is drawn. A second semicircle k with diameter MB is drawn on the same side of the line AB. Let X and Y be points on k such that the arc BX is one and a half times as long as the arc BY. The line MY intersects the line BX at C and the larger semicircle h at D. Show that Y is the midpoint of the line segment CD.

Solution

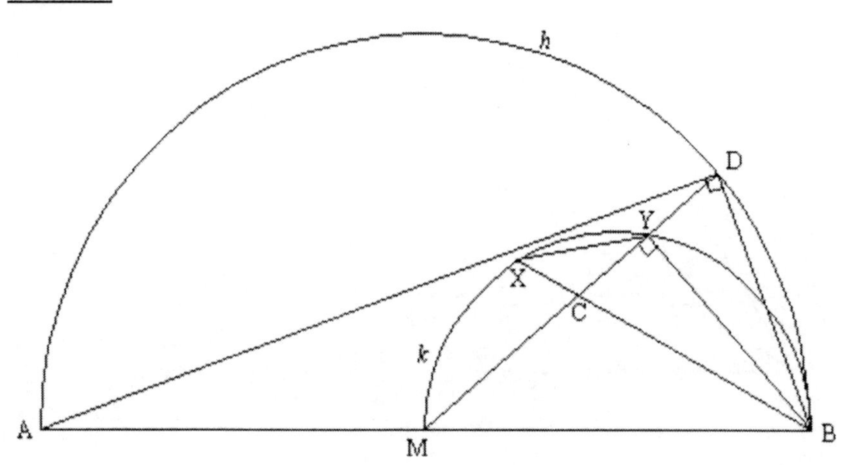

Since arc BX is one and a half that of arc BY, $\angle YMB = 2\angle XBY$, and since Y is on the semicircle k, $\angle MYB = 90°$ and $\angle YCB = 90° - \angle XBY$.

Also, since D is on the semicircle h, $\angle ADB = 90°$ and $\angle MDB = 90° - \angle ADM = 90° - \frac{1}{2}\angle YMB = 90° - \angle XBY = \angle YCB$, or triangle CBD is isosceles with CB = DB.

Therefore, BY is also the perpendicular bisector of $\angle CBD$, or Y is the midpoint of the line segment CD.

Problem 2 of Irish Mathematical Olympiad 1994

Let A, B, C be three collinear points with B between A and C. Equilateral triangles ABD, BCE, CAF are constructed with D, E on one side of the line AC and F on the opposite side. Prove that the centroids of the triangles are the vertices of an equilateral triangle. Prove that the centroid of this triangle lies on the line AC.

Solution

a) Let $a = AB$, $b = BC$ and $c = a + b$, and let J, K, R, H and G be the feet from A to BD, C to BE, E to BC, A to CF and C to AF, respectively. Also let X, Y and Z be the centroids of equilateral triangles ABD, BCE and ACF, respectively.

From X and Z draw perpendicular lines and meet CG and CK at P and Q, respectively.

Observe that $AX = GP$, $HZ = CQ$, $XP = AG = HC = ZQ = \frac{c}{2}$

and $GZ = \frac{1}{3} CG = \frac{c}{2\sqrt{3}}$, and $AX = \frac{a}{\sqrt{3}}$.

Now let $\angle PXZ = \alpha$, $\tan\alpha = \frac{PZ}{XP} = \frac{GZ - GP}{XP} = \frac{GZ - AX}{AG} = \frac{c - 2a}{\sqrt{3}c}$

Similarly, $\tan\angle QZY = \frac{c - 2a}{\sqrt{3}c}$

But $c = a + b$, or $c - 2a = 2b - c$, and
$\tan\alpha = \tan\angle QZY$, or $\alpha = \angle QZY$.

Since $XP \parallel AF$ and $ZQ \parallel FC$, XP and ZQ will intercept each other at an angle of 60°, and, therefore, XZ and YZ also intercept each other at the same angle.

With the addition of $PZ = \dfrac{c - 2a}{2\sqrt{3}} = \dfrac{2b - c}{2\sqrt{3}} = QY$ and XP = ZQ as

mentioned earlier, the two triangles XPZ and ZQY are congruent which makes XZ = ZY and thus XYZ is an equilateral triangle.

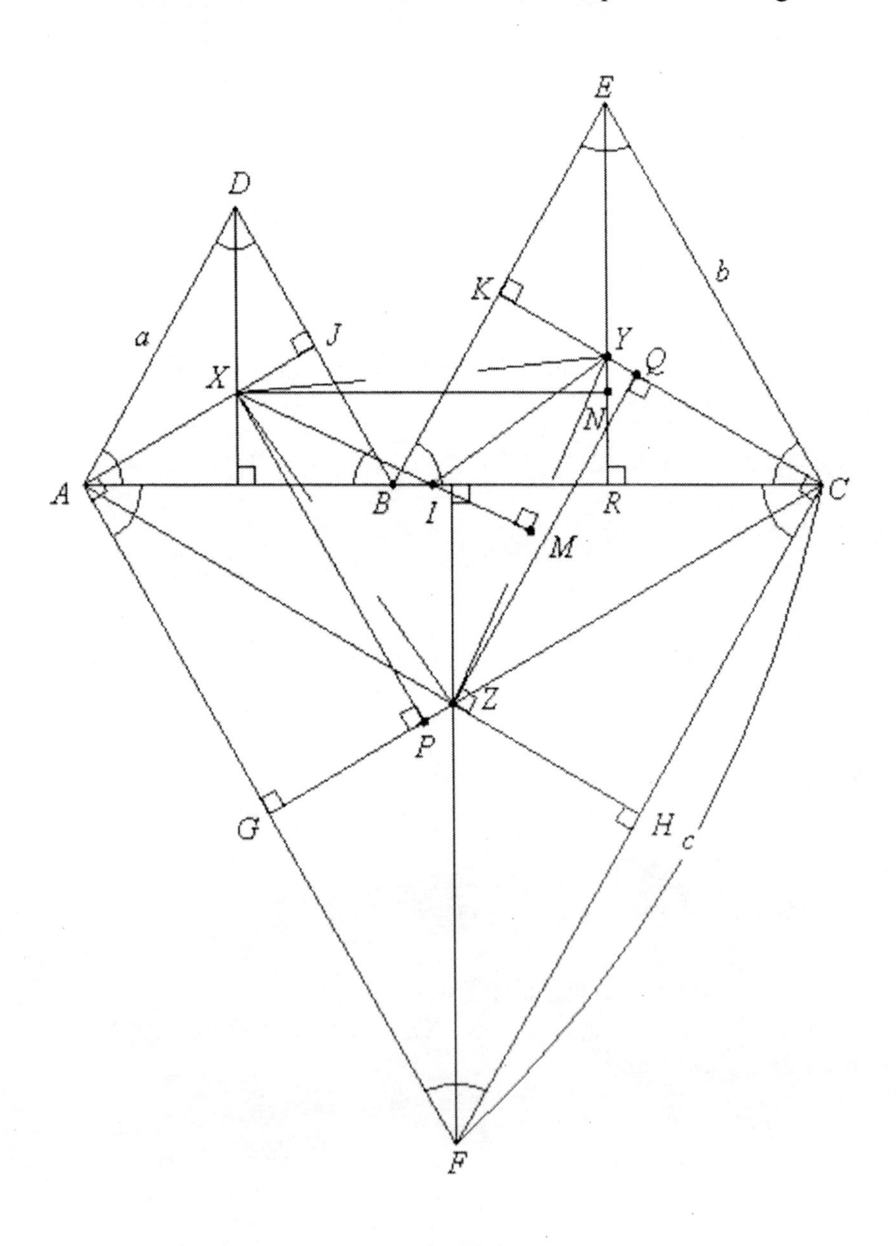

b) Now let S and M be the feet of X on AB and YZ and XM intercepts AC at I.

We have to prove that IX = IY or $IS^2 + XS^2 = IR^2 + YR^2$ (i)

But $XS^2 = \frac{a^2}{12}$, $YR^2 = \frac{b^2}{12}$, $IR = \frac{c}{2} - IS$,
and (i) becomes

$IS^2 + \frac{a^2}{12} = (\frac{c}{2} - IS)^2 + \frac{b^2}{12}$, or $IS = \frac{b+c}{6}$ is what we have to prove.

But we also have $\tan\angle SXI = \tan(60° + \alpha) = \frac{IS}{XS}$ (ii)

$$\tan\angle SXI = \tan(60° + \alpha) = \frac{\sin 60° \cos\alpha + \cos 60° \sin\alpha}{\cos 60° \cos\alpha - \sin 60° \sin\alpha}$$

However, $PZ = \frac{c - 2a}{2\sqrt{3}}$, $XP = \frac{c}{2}$

and $$\frac{\sin 60° \cos\alpha + \cos 60° \sin\alpha}{\cos 60° \cos\alpha - \sin 60° \sin\alpha} = \frac{\sqrt{3}\,XP + PZ}{XP - \sqrt{3}PZ} = \frac{2c - a}{\sqrt{3}a}$$

From (ii), $IS = XS\tan(60° + \alpha) = \frac{a}{2\sqrt{3}} \cdot \frac{2c - a}{\sqrt{3}a} = \frac{2c - a}{6}$

So now we need to prove $2c - a = b + c$, or $c = a + b$ which is obvious.

Problem 2 of Poland Mathematical Olympiad 2001

ABC is a given triangle. ABDE and ACFG are the squares drawn outside of the triangle. The points M and N are the midpoints of DG and EF, respectively. Find all the values of the ratio MN : BC.

Solution

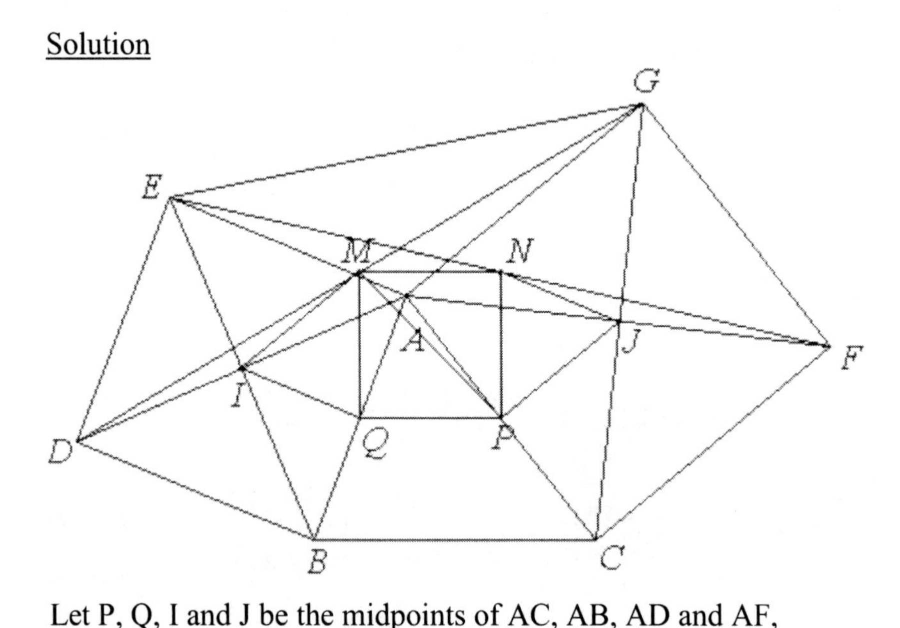

Let P, Q, I and J be the midpoints of AC, AB, AD and AF, respectively. We observe the following
QP || BC, QP = ½ BC, IQ || AE, IQ = ½ AE, NJ || AE, NJ = ½ AE, IM || AG, IM = ½ AG, JP || AG, JP = ½ AG
From there, IQ || NJ and IQ = NJ, IM || PJ and IM = PJ, ΔMIQ = ΔPJN, and we have
MQ = NP, and ∠IMQ = ∠JPN (i)
On the other hand, since IM || PJ, ∠IMP = ∠JPM.
Combining with (i), we have ∠QMP = ∠NPM, and with MQ = NP as proved earlier, MNPQ is a parallelogram. Therefore, MN || QP and MN = QP, or MN : BC = 1 : 2.
For a triangle ABC with obtuse angle BAC, the proof is similar.

Further Observation

Prove that MNPQ is a square.

Problem 3 of Balkan Mathematical Olympiad 1993

Circles C1 and C2 with centers O1 and O2, respectively, are externally tangent at point C. A circle C3 with center O touches C1 at A and C2 at B so that the centers O1, O2 lie inside C3. The common tangent to C1 and C2 at C intersects the circle C3 at K and L. If D is the midpoint of the segment KL, show that $\angle ADB = \angle O1OO2$.

Solution

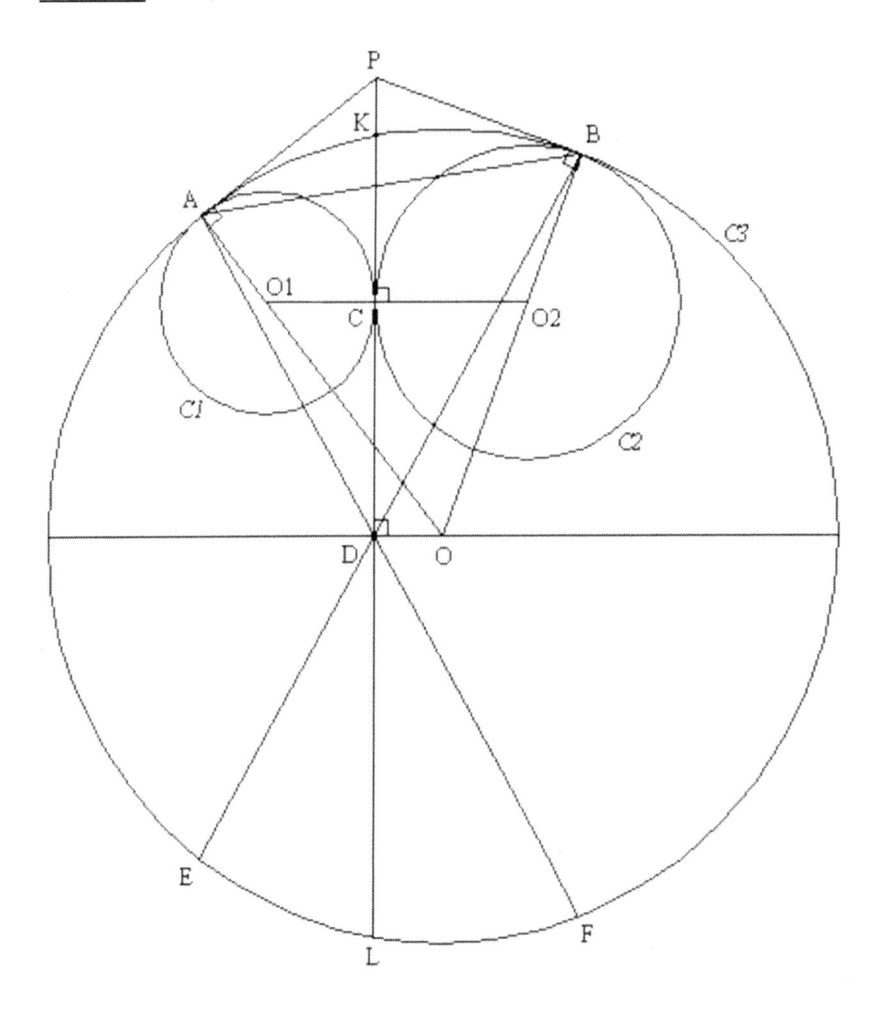

Observe that $\angle O1OO2 = \angle AOB$.

Let the tangents of C1 and C2 at A and B, respectively, meet at P. P is on the extension of LK.

Since OA $\perp$ PA, OB $\perp$ PB, and because D is midpoint of KL and O is the circumcenter, OD $\perp$ KL.

The two quadrilaterals APBO and PBOD are cyclic causing ABOD to also be cyclic on the same circle. Therefore, $\angle$AOB = $\angle$O1OO2 = $\angle$ADB.

Additional observation: From the result we have arc AB × 2 = arc AB + arc EF, or arc AB = arc EF, and since OD is on the diameter of the circumcircle, E and F are images of A and B across the diameter, respectively, which makes KL to be the angle bisector of $\angle$ADB.

CPSIA information can be obtained at www.ICGtesting.com
Printed in the USA
LVOW08s1417280414

383550LV00001B/125/P